*Life Beyond
Planet Earth?*

By the same authors

Mysterious Britain
Mazes and Labyrinths of the World
The Secret Country
A Guide to Ancient Sites in Britain
Alien Animals
Bigfoot Casebook
Earth Rites
The Evidence for Bigfoot and Other Man-Beasts
Sacred Waters
Ancient Mysteries of Britain
Modern Mysteries of Britain
Modern Mysteries of the World
Atlas of Magical Britain

Life Beyond Planet Earth?

MAN'S CONTACTS WITH SPACE PEOPLE

Janet and Colin Bord

 GraftonBooks
A Division of HarperCollins*Publishers*

GraftonBooks
A Division of HarperCollins*Publishers*
77–85 Fulham Palace Road,
Hammersmith, London W6 8JB

Published by GraftonBooks 1991

10 9 8 7 6 5 4 3 2 1

British Library Cataloguing in Publication Data
Bord, Janet
    Life beyond planet earth?: man's contacts with space
    people.
    1. Extra-terrestrial beings. Claimed observations
    I. Title   II. Bord, Colin
    001.942

ISBN 0-246-13600-6

Photoset by Rowland Phototypesetting Ltd,
Bury St Edmunds, Suffolk
Printed in Great Britain by
Butler & Tanner Ltd, Frome and London

Contents

Introduction

From the earliest times humans must have gazed at the star-studded night sky and dreamed of other worlds. Then as now, opinion was no doubt divided between those who knew that the heavens were devoid of life (Mankind naturally being the highest form of creation, and found only on Earth) and a smaller number who could imaginatively people the heavens with many worlds like our own, each with its own sentient life-forms. The desire to know if this is really so, and to make contact with these other planetary dwellers, remains as strong as ever. On a practical level it is unlikely that Mankind will ever be able to channel sufficient resources into large-scale, comprehensive interplanetary exploration, even assuming the technical barriers of time and space were overcome. Yet the dream remains, and large sums are poured into the search for signs of intelligence elsewhere in the Universe, in the hope that contact might be made. Looked at realistically, Man's efforts in this direction are 99.9% likely to be a waste of time and money . . . and yet it is that remaining 0.1%, that slight possibility of success, linked to Man's undying spark of curiosity, which drives the dreamers on. It cannot be denied that the search has produced some intriguing results, ranging from the scientifically respectable to the bizarre, and in this book we present the whole range of evidence for the possibility of extra-terrestrial life, regardless of whether or not any particular event or claim seems scientifically acceptable. And we take this opportunity to reassure readers with a non-scientific background that this book is not overburdened with scientific concepts or terminology; we hope we have presented the material in a lucid form.

It is difficult to draw conclusions from the wide variety of evidence available, especially when two authors writing a book together have to present a joint conclusion, so each of us decided on this occasion to write our own summing-up; our individual views are presented in the final chapter. But our conclusions are just as likely to be as wrong as anyone else's, and the final decision as to whether there is life beyond planet Earth remains with you, the reader!

1 Is contact feasible – or even desirable?

Although men think of themselves as rational beings, their actions both as individuals and as nations are often affected by numerous influences other than logic or rationality. If these were the guiding factors, disputes would be resolved without recourse to war, and yet there is always armed conflict occurring somewhere on this planet. Humans are in fact volatile, irrational and generally unaware of their limitations. Instead of living quietly and creatively in peace and harmony on this beautiful world, we squabble and fight and have dreams about colonizing outer space.

Anyone holding such a negative view would have considered it unlikely, even impossible, that Man would ever set foot on the Moon, and yet the impossible has happened. The dream that Man might one day travel into space has become reality. But is it possible for Mankind to achieve more? Despite all the logical arguments which can be marshalled against the likelihood of even greater achievements in space travel and colonization, Man persists in believing that ultimately nothing is impossible.

In past centuries Man's curiosity caused him to explore the oceans of the world; now in this age of the rocket motor he has turned his attention to the oceans of outer space. The expense of space travel is so great that an involuntary brake is put on the speed at which space exploration can be achieved. In addition, because of the vast distances involved, even within the Solar System, a round trip to a 'nearby' planet would take an astronaut several years to accomplish; and both these factors have caused the dream of interplanetary travel to recede. Meanwhile, it is not so expensive to explore the depths of space by other means not involving physical travel, and Man is therefore pursuing his other main preoccupation: the need to know whether we are alone.

Because we exist as an intelligent life-form, the argument goes, so must other intelligent life-forms exist elsewhere in space. That is not in fact a logical statement, but Man does not like to admit the possibility that he might be alone, a 'one-off', perhaps even an aberration. So the search for our fellow citizens in the Universe continues undaunted, despite a complete lack of success as yet on the part of the scientists.

It is not too surprising that Man's efforts to find signs of intelligent life in outer space have failed thus far, considering the difficulties involved. In the next chapter we will describe the methods used; here we concentrate on the problems facing scientists attempting to locate extra-terrestrial intelligences.

The area to be covered by the search is so vast, the problem is not to find a needle in a haystack but a needle in a billion haystacks all in one. We do not yet know the full extent of our own Solar System (there may be another planet to find, the so-called Planet X), let alone what lies beyond. Our Solar System, with the planets Mercury, Venus, Earth, Mars, Jupiter, Saturn, Uranus, Neptune and Pluto orbiting the Sun, is but a tiny part of a galaxy, the Milky Way. Our Sun is one of a hundred billion stars in this galaxy; and beyond the Milky Way are other galaxies, about 1,000,000 of them, up to 200,000 light-years in diameter, and also containing millions of stars. The most distant galaxy discovered so far, named 4C41.17 and located by astronomers in 1989, is about fifteen billion light-years away from Earth, near the edge of the visible Universe. Where do we look among all these to find intelligent life?

Scientists have virtually ruled out the other planets in our Solar System, because the conditions on them would not appear to be conducive to the development of intelligent life in the same form as we have developed. But who knows what the situation may be elsewhere in our galaxy? Perhaps some of those billions of stars have planets just like Earth circling them, where life has developed as it did here. As we shall see later, the criteria for the development of life on a planet are complex and specific, and it may be that the exact conditions required are very rarely found. We just do not know. One of the many limitations to communication with any other life-bearing planet, should it be located, is the speed of light. Using radio frequencies to communicate is a time-consuming procedure because of the time-lag: there would be

The Andromeda Galaxy (M31), about 2 million light-years away from Earth, and the nearest spiral galaxy similar to the Milky Way. The two bright spots are companion galaxies orbiting the Andromeda.

years of waiting for a reply to each message sent to a known communicant. The distances to be travelled are so vast as to be incomprehensible: our galaxy is 100,000 light-years across. A light-year is the distance that light travels in one year at the speed of 3×10^5 kilometres a second (186,282 miles per second, or 670,616,722 miles per hour), the distance in miles being approximately 5.8 trillion. Nor is the present transmitting equipment powerful enough to encompass these vast distances. The very large, thousand-metre Arecibo radio telescope in Puerto Rico could receive a signal from a similar telescope at a maximum range of a thousand light-years away. In other words scientists cannot even monitor the whole of our galaxy, let alone the galaxies beyond.

Current technology is not sufficiently advanced to enable a really effective search to be made for other inhabited worlds, and an appreciation of the relatively primitive state of current technology presents us with another reason why Mankind has had no success in finding any sign of extra-terrestrial life: he may be far too late. One current estimate is that the Universe was formed about fifteen billion years ago (one billion = one thousand million), the galaxies twelve billion years ago, and the Sun and Solar System five billion years ago, with life on Earth slowly evolving over the last four billion years. The first mammals emerged around 270 million years ago, and Man's first appearance was a mere million years ago, so his technological development has taken place during a brief eye-blink on this kind of time-scale. If other civilizations have existed in our galaxy, it is unlikely that their development paralleled ours. Intelligent beings could have evolved even before our Solar System was formed, and their development in the succeeding five billion years could place them at a level of attainment far exceeding anything humans could imagine and well beyond attempts at communication. They may even have died out several billion years ago. Even if a civilization had developed on a time-scale approximately parallel with ours, only a small disparity in time, of no more than a hundred years, could make all the difference between success and failure when it comes to locating and communicating with them. This point can be illustrated by the recent history of Western nations where, a hundred years ago, many aspects of our present life would have been unimaginable: current transport systems by road and air and the technologies of radio, television, video, computers, etc, that have been developed from the discovery of electro-magnetism. Yet it's not so long ago, and there are still people alive now who were children a hundred years ago. By the year 2100 technologists may have developed an entirely new method of communication which would make our present methods appear quite primitive. Changes take place very rapidly now that we have become a technological civilization, and the same would probably be true of a technological civilization in distant space.

In addition to this difference of development causing great problems in the locating of extra-terrestrial civilizations, it could also cause insoluble problems when we tried to understand and communicate with them. If they were any distance ahead of us, we might not be intelligent enough to comprehend them in any way; and we would be like primitive savages to them: it could be more difficult than initiating intelligent communication between man and ape.

This also presupposes that extra-terrestrial beings are the same as us physically and mentally. It is possible that their development could have followed the same course as ours; it is also possible that it did not. Although the human body seems to us to be a highly organized, efficient machine, machines can take many forms and still be efficient and effective. Terrestrial life is carbon-based, but elsewhere, in a different environment, it could be based on another element, or there could even be life not based on any element at all: 'life' could take other forms that would not be recognizable to us as life.

How would we communicate with something we could not even recognize as a living being, and what could we possibly have to communicate? It is unlikely that their motivation would be the same as ours, and this could apply to *any* space entities even if they were human in appearance. It would be unwise to base our expectations of their behaviour on what *we* would do in any given circumstance. Extra-terrestrials are as likely to be malignant and attempt to destroy us as they are to be benign and help us solve our problems.

So might it be wiser not to attempt to make contact with extra-terrestrials, and to avoid or be wary of any approaches they might make to us? There is another reason for caution: the psychological effect of contact with extra-terrestrials is unknown. The idea of extra-terrestrials among us has been made familiar through science-fiction writings and movies and, as entertainment, causes no traumas. But if it were known beyond doubt that extra-terrestrials existed, were different and might have intentions of planetary conquest, the effect on the human psyche and behaviour would be traumatic and incalculable. Dangers affecting the whole planet, like world wars, nuclear annihilation, environmental destruction, cause fear and anxiety that gnaw away at the subconscious level within humankind. But certain knowledge of the existence of extra-terrestrials would be immeasurably more dangerous to the stability of the human race.

It seems that some scientists are aware of the undesirability of tangling with extra-terrestrials. Professor Zdenek Kopal said, '. . . should we ever hear the "space-phone" ringing in the form of observational evidence which may admit of no other explanation, for God's sake let us not answer; but rather make ourselves as inconspicuous as we can to avoid attracting attention . . .'[1] Although at first contact with extra-terrestrials may seem like a good idea, the initiative could soon be taken out of Man's hands: once contact is established there is no going back.

Perhaps this nightmarish scenario is just that, and will never become a reality. We fervently hope so. In contrast, the opposite view was promulgated by many people claiming to have met extra-terrestrials in the 1950s and 1960s (see Chapters 10–12). Here the space people are beautiful beings of human appearance who are both older and wiser than humans and who possess an incredibly advanced technology. Landing on Earth in their flying saucers they make contact with certain carefully selected individuals and present a philosophy of peace, calm and brotherly love. The scientists, of course, deny that any such contact has taken place; and the fact that they are continuing with their programme of searching for extra-terrestrial signals shows their dissociation from such weird claims. But despite the problems associated with the scientifically based search, 'success' could be near. The better the equipment becomes, and the more experiments which are made, the more likely they are to have positive results. And perhaps the extra-terrestrials are actively searching for *us*, and sooner than we expect, may find us.

2 Life on Solar System planets

At present it is known that nine planets move around the Sun in almost circular orbits. The nearest to the Sun is Mercury, then come Venus, Earth, Mars, Jupiter, Saturn, Uranus, Neptune and Pluto. There are also numerous smaller bodies in the Solar System: satellites or moons orbiting some of the planets, and minor planets called asteroids, as well as comets and meteors. In terms of the Universe as a whole, our Solar System is very small and insignificant: indeed, the vastness of the Universe makes our Solar System feel positively cosy, yet even so the distances from planet to planet are difficult to comprehend. From Earth to Mars is 124 million miles (200 million kilometres), a journey which took the *Viking* spacecraft just over a year in 1976. The nearest star to us beyond our Solar System is Proxima Centauri, 4×10^{13} kilometres away, or 4.2 light-years (a light-year is a measurement of distance, not time; see previous chapter), or 24 trillion miles, an unimaginable distance for Mankind to consider travelling, and presumably equally unimaginable for extra-terrestrial beings unless they have developed a form of space travel in which our notions of time and space are meaningless.

So far, Man's attempts at exploring space have had to be confined to our Solar System, and the furthest he has got in terms of actually stepping on to alien soil is the landing of men on the Moon, our own satellite, which took place in 1969. But space probes have travelled much further into the Solar System, and have transmitted back valuable information about the other planets. Despite this great increase in our knowledge, there is still much to learn about whether life ever existed elsewhere in our Solar System, and whether it exists there now. The conditions on other planets do not seem hospitable by earthly standards, but probably we should think more flexibly when investigating the possibility of life existing elsewhere: it may be that life exists not only in our known three dimensions, but in other as yet unknown dimensions too. The bare, rock-strewn surface of Mars may be merely one aspect of that planet. If we had a key to unlock the as-yet-hidden door, we might be able to step through into another level of awareness, still Mars but a Mars transformed: not merely habitable but even a world of beauty and abundance, a kind of paradise. Such flights of fancy are usually considered unscientific, but it is worth remembering that science should be a search for truth and it does not yet have all the answers. Perhaps the scientists are asking the wrong questions: the truth often proves to be far stranger than anything ever anticipated.

But until a way is found to penetrate and explore beyond the three-dimensional Universe, we must be content with the results of the scientists' exploratory work. When studying the Solar System, and especially the other planets, scientists have always endeavoured to establish whether any form of life has ever existed there, and some intriguing findings have come to light in recent decades. So visiting the planets in order, beginning nearest the Sun, we will briefly describe what sort of environments they offer, and whether any signs of life have been found.

Mercury

During the last century it was believed that there was a planet, known as Vulcan, orbiting even closer to the Sun than Mercury, but today Vulcan's existence is doubted – unless it is a large asteroid within

The surface of Mercury photographed from a distance of 21,700 miles (35,000 kilometres) by Mariner 10 in 1974. The large crater is about 50 miles (80 kilometres) in diameter.

Mercury's orbit. So we will begin with Mercury, which is a small planet with a rocky surface, lacking an atmosphere and heated by the Sun's radiation so that the surface rocks reach a temperature of 300°C and more, and drop to −170°C during the three-month night. Mercury's cratered surface looks somewhat like our Moon's, and, not surprisingly in view of its extremes of temperature, no signs of life have been found. At the end of the last century some astronomers claimed to see a network of streaks and lines on the planet's surface, similar to the 'canals' noted on Mars. The discovery was reported in *Scientific American* as follows:

> [Giovanni] Schiaparelli has discovered many marks upon the disk of the planet which had not been noticed before, and he has made a little map or diagram which shows that these marks strikingly resemble some of the features discovered in recent years on Mars. They are elongated streaks running in various directions, and frequently presenting at their points of junction the appearance of an enlargement or knot. Similar streaks on Mars have been assumed to be long narrow seas or water-courses. The geometrical figures formed by the intersection of these streaks on Mercury strikingly resemble those on Mars. In one place there is a shape of this kind that roughly resembles a huge figure 5, covering a quarter of one hemisphere of the planet.[1]

But the recent close observations achieved by spacecraft have not confirmed these findings, and it is probable that the lines were an illusion, for not all the astronomers who looked for them were able to see them.

Venus

Although not so close to the Sun as Mercury, Venus is also a very hot place to live: the surface temperature is 475°C. The atmosphere consists largely of carbon dioxide with only traces of oxygen, contrasting with our own oxygen-rich atmosphere with only traces of carbon dioxide. The atmosphere is also a hundred times thicker than Earth's, and the planet has a continuous sulphuric acid cloud cover. Infra-red radiation is trapped by the atmosphere with the result that the 'greenhouse effect' is rampant, and Venus is apparently not hospitable to any form of intelligent (or indeed any other) life. There may have been oceans on Venus in the past, but any water has long since evaporated and been lost. When, in the distant future, the Sun nears the end of its life and increases its heat output as it becomes a red giant star, the Earth will get hotter and eventually become like Venus is now.

Our first information on Venus obtained by space probe came from *Mariner 2*, which was launched by the Americans in 1962 and flew past the planet, followed in 1967 by *Mariner 5* and in 1974 by *Mariner 10*. In 1970 the Russians landed a probe, *Venera 7*, on the Venusian surface, followed by others in 1972 and 1975: all lasted only a short while before succumbing to the hostile environment, but a few pictures were sent back to Earth. In 1978 Russian and American probes landed and were able to send back much more information. But the space probes did not confirm the existence of lines in the form of a wheel with spokes, which some astronomers have recorded. These are similar to the lines seen on Mars and Mercury and equally mysterious. One late nineteenth-century astronomer who saw them was A. E. Douglass, who wrote:

> On the day succeeding my first good view, I spent nearly the whole afternoon without catching a single certain glimpse. Suddenly the seeing improved for an instant, and I saw the same markings unmistakably. If it had not been for that glimpse I would have gone away perfectly ready to believe that no markings existed. I am not surprised that other astronomers doubt them.[2]

The lines may simply be optical illusions, or, as more recent research has suggested, they may be cloud patterns caused by ultra-violet markings. But it is certainly unlikely that they have any connection with the presence of life on the planet, despite the regular nature of the markings. There have, of course, been claims that humanoid entities live on Venus (see later chapters), but so far no scientific confirmation of these claims has been forthcoming.

The Moon

Earth's satellite is one-quarter the size of Earth: a dry, airless, rocky

The full Moon, photographed from the Apollo 11 *spacecraft in 1969.*

place appearing totally uninviting as a place to live. The so-called 'seas' (*maria*) are dry rock plains, not oceans as the early astronomers thought; and the scientific research done into rock and soil samples brought back by astronauts and unmanned space probes has shown that the Moon does not support life. But still there are mysteries. Some relate to the Moon's formation: was it originally a part of the Earth which broke away; or was it a separate planet which was captured by the Earth's gravitational field when it came too close? Some relate to inexplicable happenings on the Moon, like the bright, star-like points of light seen on its surface, either in the form of brief flashes or a steady light lasting just seconds or an hour or two. These may be sunlight

Human astronaut Eugene A. Cernan, commander of the Apollo 17 spacecraft, testing the Lunar Roving Vehicle during a Moon landing in 1972.

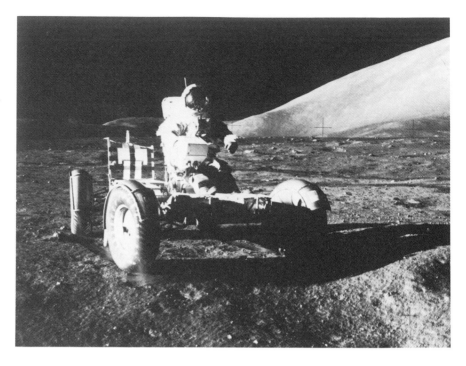

reflections, or caused by lava and gases, or some kind of moonquake lights – but as yet they remain imperfectly understood.

Just one example of these anomalous lights, from the many that have been recorded, was the flashing watched for one and a half hours on 17 January 1956 by Robert Miles in California:

> I noticed a flash of white light that caught my eye. At first I thought it could have been a lunar meteor. But it kept flashing on and off . . . The light was very bright but changed its color to a very bright blue, like an arc light. It was brighter than the sunlit portions I was looking at.[3]

Various anomalous spots, streaks, mists and dust clouds have also been seen, as well as 'canals' similar to those noted on Mercury, Venus and Mars. Some of these linear features are likely to be optical illusions, some may be real but representing rock fractures or other natural surface markings, rather than being evidence of intelligent beings at work.

However, even though extra-terrestrials may not have dug canals on the Moon, they could have established bases there, just as Earthmen are hoping to in the not-too-distant future. Three people who believe that this has happened are George Leonard (who wrote *Somebody Else is on the Moon*), Fred Steckling (who wrote *We Discovered Alien Bases on the Moon*), and Don Wilson (who wrote *Our Mysterious Spaceship Moon* and *Secrets of our Spaceship Moon*). By studying NASA close-up photographs of the Moon's surface, George Leonard was able to locate 'manufactured vehicles' and 'super rigs' at work in the craters, as well as ground markings, insignia and high-rise signals, some of these markings and/or structures resembling letters of the alphabet, runes, arrows, antennae and towers. Leonard's hypothesis is that perhaps the extra-terrestrials are mining metals, or even repairing the Moon which they guided to its present location aeons ago. The extra-terrestrials may need to visit Earth periodically, he suggests, for water, animals and other necessities, and it is on record that UFO occupants have been seen outside their landed craft, examining plants, taking water and even stealing animals. Leonard also suggests that the Americans undertook

Two objects which have rolled across a Moon crater. The tracks are 900 and 1,200 feet (273 and 365 metres) long and may have been caused by moving boulders, though George Leonard (in Somebody Else is on the Moon) *feels the objects do not look like boulders, and comments on the 'tread mark' of one trail, as well as saying that the smaller object 'came up out of a crater' before rolling downhill.*

their Moon exploration programme because they knew extra-terrestrials were up there and wanted to find out more about what was going on, while keeping the truth hidden from Earth people.

Don Wilson also believes that the authorities are not telling all they know, and in his books he lists various Moon mysteries to support his idea that the satellite is in fact hollow, having been converted into a spacecraft by aliens whose own ship was damaged. Fred Steckling also relies on close examination of NASA photographs to locate strange artefacts suggesting the presence of alien beings on the Moon. However, photographic interpretation of this kind is dangerous because the mind naturally strives to make intelligible any jumble of data which the eye presents to it, and will tend to see regular patterns and shapes which are in fact caused by shadows, brightly lit areas, and enlarged grain in the photographic film. The process at work is similar to that which caused astronomers to see linear features, 'canals', on the surfaces of Mercury, Venus, Mars and the Moon. There is also plenty of scientific evidence that counters the hypotheses formulated by these authors; for example, data from the moonquake recorders left on the Moon by the *Apollo* astronauts prove that the Moon cannot be hollow, as alleged by Don Wilson. However, the use of the Moon as a base by extra-terrestrials seems a feasible idea if they are indeed travelling in our Solar System; and maybe some of the UFOs sighted on Earth are hopping across the 222,500 miles (384,000 kilometres) that separate us from our satellite.

Mars

The most likely planet in our Solar System where life might be found is Mars, the planet next to Earth and similar in some respects, though it is cold: the warmest temperature achieved is 27°C, the average is below freezing, and the night-time temperature is very cold. The existence of liquid water seems unlikely in these conditions, but some surface features resemble dry river beds, as if there was once running water on the planet, and some scientists believe that oceans have covered large areas of the planet and then evaporated, the process being repeated after millions of years. There may be water in the form of ice below the surface, which could be tapped by any future colonizers of the planet.

The atmosphere, largely carbon dioxide, is thin, and at the poles there is frozen carbon dioxide, which means the ground temperature must be as low as $-125°C$. Nor is there sufficient ozone in the atmosphere to protect the surface from the Sun's ultra-violet radiation. So the conditions on Mars today do not seem very conducive to the development of intelligent life. Yet rumours of life on Mars have persisted, and recent evidence brought back by space probes has intensified the speculation. America's unmanned *Viking 1* spacecraft put a 'lander' on the surface of Mars on 20 July 1976, followed by one from *Viking 2* on 3 September, and they took photographs of the planet's surface, tested the atmosphere, and collected soil samples to bring back to Earth. Experiments conducted on them were inconclusive: positive evidence of life was not found – but there may be life-forms in the soil of a type that did not respond to the specific experiments that were carried out.[4] Scientists are also looking for fossil micro-organisms in Martian rocks. A meteorite found in Antarctica in 1979 and believed to have almost certainly come from Mars was found to contain 'high concentrations of organic materials', leading to renewed enthusiasm for the search for life on the planet.[5]

Apart from the unsolved mystery of whether life in any form exists or has existed on Mars, there are plenty of other strange aspects to the planet, like the 'canals' which were discovered by Giovanni Schiaparelli in the late nineteenth century. Some astronomers could see them, others could not; but it was Percival Lowell of Flagstaff, Arizona, who drew the most detailed maps of the canal network. However, the photographs transmitted by *Mariner* and *Viking* spacecraft have proved conclusively that the canals do not exist – at least as surface features. There are still occasional sightings of lines corresponding to Lowell's canals and one possible explanation is that they are atmospheric features which come and go.

Slight evidence for the existence of vegetation on Mars, obtained by spectroscope in 1957, has been dismissed as unlikely, but in the early 1980s two scientists showed two photographs of a Martian rock taken years apart by a camera on one of the landed probes. The photographs showed greenish patches which had changed over the years, and a spectral analysis was similar to that obtained from lichen-covered rocks on Earth.[6]

Another mystery is the occasional sighting of luminous phenomena in the form of flares and light flashes. In December 1900 the Lowell Observatory in Arizona reported a shaft of light lasting seventy minutes, though the usual life-time is five seconds to five minutes. These so-called lights may be white vapour clouds, or volcanic eruptions, rather than signs of intelligent life.[7]

Also probably not really mysterious are the origins of the two Martian satellites, Phobos and Deimos. Both moons are small, and were not observed from Earth until 1877 when astronomer Asaph Hall saw them (although in 1726 Jonathan Swift described them in *Gulliver's Travels*, and no one knows how he could have known the details so accurately). It has been suggested that Phobos is hollow and an artificial satellite, but scientists do not accept this theory.

This photograph of the surface of Mars was taken in July 1976 from the Viking 1 *orbiter, and shows the rock formation which resembles a human face. Pyramidal structures are also visible. (The speckles were caused by computer error.)*

Nor are they much interested in the current excitement which has grown up around photographs showing apparently artificial structures like pyramids and giant faces on the Martian surface. The pyramidal structures are certainly puzzling, but may have been caused naturally by wind and sand erosion.[8] It is the giant face in the Cydonia region which has created the greatest controversy. It first showed up on a photograph produced by one of the *Viking* craft in 1976. Close by is a particularly fine 'pyramid', as well as other 'structures' classed by researchers as a 'city'. Sceptics shrug off the 'face', saying it is nothing more than a trick of light and shadow, but Mark Carlotto, an optical

engineering expert, has used computer enhancement on the photographs showing the face and his results have convinced him that the features were carved by 'intelligent design': 'The image enhancement results indicate that a second eye socket may be present on the right, shadowed side of the face; fine structure in the mouth suggests teeth are apparent.'[9] Researchers have also claimed to see a helmet or head-dress, cheek ornaments, and an indentation over the right eye.

The rock bearing the 'face' is a mile (1.6 kilometres) long, and it is obvious that such an undertaking would have involved considerable labour. Perhaps it was carved as a signal from the Martians to other intelligences in the Solar System, a signal saying 'There are people living here.' However, even if this were the case, it may not mean that the Martians still inhabit Mars: they may have died out thousands or millions of years ago. In addition to the other structures located, researcher Richard Hoagland claims that a line drawn from the centre of the 'city' across the face to a massif or cliff marks the position of the Sun at the moment of the Mars summer solstice 500,000 years ago, so if intelligent beings did live on Mars, it is possible that they did so that long ago.[10] Throughout the twentieth century science-fiction writers have regaled us with tales of Martians, and in recent years some of the people who have claimed encounters with extra-terrestrials have identified them as coming from Mars. So despite the inhospitable appearance of the red planet's stony surface, perhaps after all it has been home to intelligent beings – and maybe still is . . .

Jupiter, Saturn, Uranus and Neptune

Of the four giant planets, Jupiter is nearest to Earth. It has a strong magnetic field, ten times stronger than Earth's, and emits intense radio waves. There is no solid surface to the planet, and like the other giant planets it is cloaked in thick gases. All four have primitive atmospheres, the same as when they were first formed 4.6 billion years ago. But this apparently hostile environment does not necessarily mean that primitive life has not formed; something may be at home there, perhaps giant creatures drifting in the atmosphere. Jupiter has fourteen satellites and a system of rings, the latter discovered by the *Voyager* spacecraft in 1979. The four largest moons have been named Io, Europa, Ganymede and Callisto, and they all have solid surfaces. Io has active volcanoes, Europa has a network of lines in its icy shell, and Ganymede is criss-crossed by systems of grooves running parallel for 6–620 miles (10–1000 kilometres). Saturn, Uranus and Neptune, like Jupiter, have systems of rings and numerous satellites (Saturn ten, Uranus five, Neptune at least six). Both Saturn and Uranus were bypassed by *Voyager 2* on its long journey through the outer Solar System, and thousands of spectacular close-up photographs were sent back to Earth.

Jupiter with its four planet-sized moons, in a montage prepared by NASA which shows them not to scale but in their relative positions.

Saturn's largest moon is Titan, also the second largest moon in the Solar System. In 1989 it passed in front of a bright star, and this occultation was studied by astronomers who were surprised to discover that its atmosphere is made up almost entirely of nitrogen. This means that Titan's atmosphere is very similar to that of Earth more than 500 million years ago, before oxygen started to form here. We should know more about Titan after 2000, in which year a European space probe will be sent to land on the surface. It is unlikely to prove habitable, for the

Saturn and its moons, in a NASA montage using photographs taken by Voyager 1 *in 1980. Titan in its distant orbit is at the top right, and the foreground moon (in front of Saturn) is Dione.*

temperatures are very low; two of Saturn's other moons, Enceladus and Tethys, consist mainly of ice.[11]

In August 1989 *Voyager 2* passed within 3,000 miles (4,800 kilometres) of Neptune, a planet whose existence has only been known since 1846. Its diameter is four times that of Earth and it has two large moons, Triton and Nereid, and at least four others. It is thirty times as far as Earth is from the Sun, and it takes 165 years to orbit the Sun, so winter on Neptune lasts more than forty years. It is *very* cold, and like the other gas giants, Neptune is almost certainly inhospitable to intelligent life.

Pluto and Planet X

According to present knowledge, Pluto is the 'last' planet in the Solar System, and furthest from the Sun, though sometimes, because of its eccentric orbit, it passes closer to the Sun than does Neptune. Pluto's existence was not known for certain until 1930, though it was suspected. It has a moon named Charon, and an icy methane environment. It has been suggested that Pluto was once a satellite of Neptune's, broken in two by an encounter with another planet, both Pluto and Charon being expelled to the edge of the Solar System. The planet's eccentric orbit may hark back to this event. The intruding planet which was responsible was the mysterious Planet X, whose existence may also be the reason why Neptune's moon Triton orbits in the opposite direction to other large moons. Both Neptune and Uranus also have irregularities in their orbits, again possibly caused by Planet X. If it exists, it is calculated to be 20,000 times fainter than the Pole Star when observed from Earth, but four times bigger than the Earth.

3 Man sends messages into space – and receives some

Science tells us that life on Earth depends on energy radiated from the Sun, the star around which all the planets in our Solar System are in orbit. There are about 100,000 million stars in the Milky Way galaxy, and it is logical to assume that around many of these are planets in orbit. The current estimation is that there are about six million million million planets like Earth orbiting stars like our Sun in the whole of the Universe, that is the Milky Way galaxy and all the other galaxies beyond. But even if the first essential of a planet orbiting a star is achieved, there are still many criteria to be fulfilled before intelligent life can exist, or even life of any kind. The vast distances involved have made it exceedingly difficult for scientists to search for life outside our Solar System: light takes a 'mere' eight hours to cross the Solar System, but 100,000 years to cross the galaxy. So in effect we can learn very little about other solar systems within our own galaxy, and even less about those in other galaxies.

Three techniques are in use by scientists looking for the planets of other stars: searching for them directly through telescopes (only really feasible by means of orbiting space telescopes); measuring changes in a star's velocity by examining its spectrum (change suggests that the gravitational pull of a planet is tugging at the star); recording and observing the star's movement and deducing the existence of planets if it does not follow a straight line through space. When these methods are used, some stars have been located which appear to have planets accompanying them. One is Barnard's star, six light-years away from us, and the nearest star in the northern equatorial hemisphere. Its course across the sky over the last sixty years has been photographically recorded, and slight deviations from a smooth track noted. The deviations can be explained by supposing that two planets orbit the star, one with a mass like that of Jupiter, the other with a mass half that of Jupiter. Other stars which are known to be comparable to our Sun and therefore might have planets in orbit around them are Alpha Centauri, Epsilon Eridani and Tau Ceti, but data on these stars are difficult to obtain because of their vast distances from Earth. Alpha Centauri is 4.24 light-years away and Tau Ceti even further at twelve light-years.

Searching for planets orbiting distant stars is one way of investigating the possibility that life has developed outside our Solar System. Another way is examining interstellar absorption spectra and comparing them with the spectra of simple life-forms like bacteria. Some scientists believe there is no positive evidence yet obtained by this method of research; others disagree. Two astronomers who support the notion of interstellar bacteria are Fred Hoyle and Chandra Wickramasinghe, who have concluded that 'the facts of astronomy point strongly to interstellar space being chock-a-block with biological material'.[1] They have written some interesting books about their work, including *Cosmic Life-Force* and *Lifecloud*. In recent years more evidence has been found to support their theories. More scientists now accept that organic material is present in meteorites, and that terrestrial contamination is not always a satisfactory explanation for this. It is even being seriously suggested that life on Earth developed following the bombardment of the planet by life-bearing comets.[2]

Search for Extra-Terrestrial Intelligence (SETI)

Before the discovery of radio-waves, Man was very restricted in the methods he could employ when he wanted to attempt to make contact with extra-terrestrial civilizations. Today the ideas of a hundred years ago and later appear ineffectual if not ludicrous. How will today's efforts appear in a hundred years' time when radio has possibly been superseded as a means of communication? Who would now suggest communicating with the Martians by planting broad strips of forest in a giant right-angled triangle in Siberia and sowing wheat within that shape, so that the interior colour would be all the same? The German mathematician Karl Friedrich Gauss suggested this in the early nineteenth century; and another idea was to plant squares on all sides of the triangle, to form the Pythagorean theorem. Also seriously suggested was that canals should be dug in the Sahara desert, forming geometric figures with sides 20 miles (32 kilometres) long, then filling the canals with water and at night spreading kerosene on the water and setting fire to it. This was suggested by the astronomer Joseph Johann von Littrow; Frenchman Charles Cros wanted the government to build a vast mirror to reflect sunlight to Mars. As a variant on this, it was suggested that several mirrors could be used to transmit semaphore. Two other suggestions from the end of the last century were to use electric lamps to illuminate several square miles of snow in the shape of a human figure, which would be visible from outer space; and to locate two lines of electric lights in a cross shape on Lake Michigan, the lights timed to flash on for ten minutes and then off for ten minutes. Even as recently as 1941 Sir James Jeans suggested that during Mars' close approach to Earth that year, searchlights could be used to flash the prime numbers towards the red planet.

Fifty years on, scientists are using far more sophisticated means of searching for and attempting to communicate with extra-terrestrial intelligences. But at the same time the naïve attitude of 1941, that so simple a method as flashing searchlights might alert Martians to the existence of life on Earth, has been replaced by a realization that a serious attempt at interplanetary communication is blocked by so many problems as to make the attempt almost not worth the effort and expense. 'Almost' . . . yet despite the problems scientists and governments continue to expend time and money in the faint hope that one day – soon – they will be able to speak to, or at the very least listen in on, alien intelligences. Some of the problems to be faced are:

Where should the search be concentrated? At which of the many millions of stars?

What frequencies of the billions along the technically vast electromagnetic spectrum should we search? Which might *they* be using?

Would we be able to understand any signal or message we might pick up? And what message should we transmit into space?

How can the incredible time-delay factor be overcome, remembering that light-waves and radio-waves travel at a 'mere' 186,282 miles (300,000 kilometres) per second, and that communication with distant stars could take years for each question and answer to travel back and forth?

Ought we to be utilizing other methods than radio – for example, tachyons (theoretical particles travelling faster than light) or laser beams?

These questions indicate some of the difficulties inherent in the search for extra-terrestrial intelligences.

But since radio-waves were discovered in the last quarter of the nineteenth century, scientists have been employing them in schemes designed to locate and communicate with the occupants of other planets, beginning at the turn of the century with Nikola Tesla, the Croatian-American scientist (1856–1943), who was a pioneer in the fields of electricity and radio. However it was not until the 1960s that the search for extra-terrestrial radio signals began in earnest. At the beginning of that decade, radio-astronomer Frank Drake began his programme code-named Ozma, using an 85-foot (25-metre) radio-telescope at Green Bank, West Virginia, to search for signals coming to Earth from outer space. He observed two nearby stars, Tau Ceti and Epsilon Eridani, for 200 hours spread over three months, but no intelligible signals were picked up. There were some false alarms, however: one day a regularly pulsing signal was located when the telescope was facing Epsilon Eridani, and it faded when the antenna was moved away from that star. It could not be located again but was later picked up by another receiver not pointing at Epsilon Eridani. Very probably the signal originated from an aeroplane, as on other occasions spurious signals were caused by lorries and planes.

There have been numerous similar experiments to Project Ozma in the thirty years since, and especially after 1970. Various celestial bodies have been targeted: from as few as two or three stars, to an all-directional search, but the results have been negligible. In 1976 Benjamin Zuckerman and Patrick Palmer spent 500 hours spread over four years at Green Bank Observatory surveying nearly 700 stars, choosing the ones thought most likely to have life-bearing planets, and although they were using equipment much more sensitive than that used for Project Ozma, they found nothing to indicate the presence of intelligent life. SETI programmes continue to be developed, using the most modern electronic and computer technology available. Paul Horowitz has run a SETI programme at the Harvard-Smithsonian Observatory in Massachusetts since 1983. Code-named META (Mega Channel

Extra-terrestrial Assay), the programme scans 8.4 million channels simultaneously and the entire northern sky is covered in six months. Despite such great leaps forward, no positive results have yet been achieved, and the likelihood of locating intelligent signs remains uncertain.

Yet the scientists are not down-hearted, and early in 1990 it was decided to devote even more money to the search, when NASA approved the spending of $100 million on a ten-year project. A radio receiver capable of scanning fourteen million channels of radio-waves simultaneously is to be built at the SETI Institute in Mountain View, California, with the switch-on scheduled for October 1992. The receiver will sift through information coming from radio-telescopes, and the search will be two-fold: one covering the whole sky, the other concentrating on the 770 stars nearest the Sun, that is those up to eighty million light-years away. Dr John Billingham of NASA is confident of success: 'After looking at the quantity of potential life sites in the Universe, it seems absurd to believe that life would have only come forth here on Earth.' He also feels the effort is worth the money: 'I believe that it would have the same ultimate impact on our understanding of the Universe as the Copernican revolution. In this case we would be finding that we might not be the highest order of intelligence in the Universe.'[3] As we suggested in Chapter 1, such a discovery would have a traumatic effect on mankind, and this effect is being seriously underestimated by the scientists so eager to achieve a scientific breakthrough regardless of its effects. But when the enormity of the task is considered, $100 million is a drop in the ocean, and we do not expect to see any positive results.

There *are* temporary excitements, however, like the 'WOW' signal picked up in 1977 by the Ohio State University Radio Observatory (so-called because the technician wrote 'WOW' on the recording tape), but the intense signal was brief, and never repeated. It could not be correlated with any man-made or natural radio sources.[4]

The numerous false alarms have had various causes: aircraft, terrestrial radar signals, Earth satellites, the regular pulses of radiation sent out by pulsars (radio stars) and quasars (quasi-stellar radio sources). After the initial excitement, the astronomers receiving anomalous signals are usually able to track them down to their mundane sources.

But occasionally there is no obvious explanation to be found; or else, as in the following case, the events took place too long ago for modern scientists to be able to pinpoint the likely source with absolute certainty. Nikola Tesla, whose early exploits in electricity and radio have already been mentioned, picked up some strange signals in 1899 on his powerful radio receiver that he had constructed on Knob Hill near Colorado Springs, Colorado. As he later reported:

> The changes I noted were taking place periodically, and with such a clear suggestion of number and order that they were not traceable to any cause then known to me. I was familiar, of course, with such electrical disturbances as are produced by the sun, Aurora Borealis and earth currents, and I was as sure as I could be of any fact that these variations were due to none of these causes . . . It was some time afterward when the thought flashed upon my mind that the disturbances I had observed might be due to intelligent control . . . The feeling is constantly growing on me that I had been the first to hear the greeting of one planet to another.

Tesla personally believed that the signals came from Mars, though it is now thought possible that he was actually picking up the natural radio-waves from pulsars or similar stellar objects.[5] Italian physicist Guglielmo Marconi, also a radio pioneer, was involved in transatlantic signalling before the First World War, and in 1920 he reported that his stations had been hearing strange signals for some years. The signals seemed to be in code, but were apparently meaningless; though Marconi believed they originated on Mars.

It was during the 1920s that excitement about Mars reached a peak. The planet was due to come very close to Earth in 1924, and there were great hopes that communication with the Martians could be established, perhaps by sending out high-powered searchlight beams or by picking up signals. Freak signals were indeed received, as reported in the *New York Times* during late August 1924. The following excerpts from these press reports give details of the most intriguing events.

> An attempt by British wireless experts to 'listen in' on Mars resulted in strange noises being heard at one o'clock this morning. The source of the noises could not be ascertained by the experts.

The attempt was made on a twenty-four tube set erected on a hill at Dulwich. Representatives of the Marconi Company and of London universities were present.

Tuning in started at 12.30 A.M., and at 1 A.M., on a 30,000-metre radius, sounds were heard which could not be identified as coming from any earthly station.

The sounds were likened to harsh dots, but they could not be interpreted as any known code. The noises continued on and off for three minutes in groups of four and five dots.[6]

*   *   *

The regular signals blotting out other messages, which have led radio experts here seriously to consider the theory that Mars is trying to 'tune in', were received at the Point Grey wireless station again today and also were heard by the wireless expert at the Merchants Exchange.

'The signal has been noticed at the same hour practically every day for four weeks or more,' declared C. W. Mellish, wireless operator at the government station at Point Grey. 'It is absolutely distinctive and cannot be attributed to any known instrument, or to static or to leaking transformers in Vancouver.'

This morning the signal, which dominated the air lines, was heard at 5.20 and at 7.12, at the same time to the minute that they came in on the previous days. It again came in four groups of four dashes or rather four 'slepps' so powerful that they could not be 'tuned out'.

The sounds had not been considered seriously by the operators until the last day or two, Mr Mellish stated.[7]

*   *   *

The development of a photographic film record of radio signals during a period of about twenty-nine hours, while Mars was closest to the earth, has deepened the mystery of the dots and dashes reported heard at the same time by widely separated operators of powerful stations.

C. Francis Jenkins of Washington, inventor of the device, which he calls the 'radio photo message continuous transmission machine', was induced by Dr David Todd, professor emeritus of astronomy of

Amherst and organizer of the international 'listening-in' for signals from Mars, to take the record.

The film, thirty feet long and six inches wide, discloses in black on white a fairly regular arrangement of dots and dashes along one side, but on the other side at almost evenly spaced intervals are curiously jumbled groups each taking the form of a crudely drawn face.

'I don't think the results have anything to do with Mars,' says Mr Jenkins. 'Quite likely the sounds recorded are the result of heterodyning or interference of radio signals. The film shows a repetition, at intervals of about a half-hour, of what appears to be a man's face. It's a freak which we can't explain.'[8]

Later in the 1920s, scientists were puzzled when they picked up long-delayed echoes (LDEs) of short-wave radio signals. Records were kept of the LDEs by Dr B. Van der Pol in the Netherlands, and many years later, in the early 1970s, Scottish science writer Duncan Lunan began an intensive study of these records. By plotting the echoes he produced a map of the stars of the constellation Boötes, and stated his theory that there might be an alien space probe orbiting the Earth which was from Epsilon Boötis and responsible for the LDEs. Various criticisms of this theory, and especially of the map, were voiced, and attempts made to explain the LDEs as natural phenomena. Lunan himself later accepted that new evidence ruled out Epsilon Boötis as the origin of his hypothetical probe; yet the mystery of the LDEs still remains.[9]

It seems that every so often there is an announcement of anomalous radio signals heralding the discovery of intelligent life elsewhere in the Universe. Yet these events invariably seem to be quickly forgotten, once they have been thoroughly investigated and found to have probably resulted from some natural cause. The truth is that the search for extra-terrestrial intelligence has so far 'only scratched the surface': those are the words of Michael J. Klein, who is in charge of NASA's SETI project. New equipment being brought in in the late 1980s and 1990s will be able to listen to tens of millions of channels at a time, and a computer will monitor the signals for signs of intelligent life. But even with the new equipment, the task facing the SETI project members is an immense one, as is admitted by Frank Drake, who has been involved with SETI throughout its life, and who himself headed Project Ozma thirty years ago. He commented, 'Many people think the project should succeed in the next year or two. That's probably not going to happen. The Universe is so huge and the cosmic haystack so large, it will take us decades. The consequences of success are so great that it's worth dealing with the problems.'[10]

Messages sent from Earth into space

Even if Mankind were able to make contact with extra-terrestrial intelligences, would we be able to communicate in any meaningful way? Would we have a common language? It is unlikely that extra-terrestrials speak English, or any other terrestrial language – but, having said that, see Chapters 6–13 where there are many reports of extra-terrestrials doing just that! It is assumed that they will be able to understand the language of mathematics, and also the laws of physics and chemistry, which are thought to be universal and unchanging.

Messages based on these assumptions have already been launched into space, beginning in 1972 and 1973 with the spacecraft *Pioneer 10* and *11*. They were sent to Jupiter, and *Pioneer 11* was then sent on to Saturn, both craft continuing on through the galaxy. They carry identical engraved plaques describing where they come from and the people who sent them. Of course this information will be useless to extra-terrestrials if they do not have eyes, or cannot see in the way that we see.

But assuming that they can see the plaques, they will still need to be able to decipher the information: it shows the position of our Solar System in the galaxy and of Earth within the Solar System. It also shows a male and a female human being, the man having one hand raised in a gesture of friendship. It has since been pointed out that this could equally well be interpreted as a rude or belligerent gesture, and any alien reading it might believe that Earthmen were intent on warfare. But we need not worry, for as is made clear by Donald Goldsmith and Tobias Owen in their book *The Search for Life in the Universe*, the *Pioneer* spacecraft are unlikely to encounter any star systems on their journey, and if they did, it would not be for at least 100,000 years and to encounter a planetary system like ours might take ten billion billion years – by which time the plaques will be worn away by interstellar dust.[11] So the *Pioneer* message is a symbolic gesture rather than a serious attempt at communication.

The first intentionally transmitted radio message to the stars was broadcast on 16 November 1974, by the 1,000-foot (305-metre) diameter radio-telescope at Arecibo, Puerto Rico. The director of the observatory at that time was Professor Frank Drake, and the event celebrated the renovation of the telescope. The message was aimed at the star cluster M13, consisting of 300,000 stars, at a distance of 24,000 light-years from Earth, and will take 25,000 years to get there. So we cannot expect a reply in our lifetime. The message, in the form of 1,679 pulses, gives a framework with 23 characters across and 73 characters down and the information given is: one to ten in binary arithmetic; the atomic numbers of hydrogen, carbon, nitrogen, oxygen and phosphorus to show that life on Earth is composed mainly of these elements;

The message transmitted to star cluster M13 using the Arecibo telescope in 1974.

the DNA double helix; the shape of a human being with its average height; the Earth's population shown as four thousand million; the Sun and its nine planets with Earth displaced to show its importance; and the Arecibo telescope dish with its diameter.

Although the Arecibo message was the first to be deliberately sent out, there has in fact been a continuous leakage of microwave beams from radar and television transmitters since those systems came into use on Earth, and by now these signals will be able to be picked up forty to fifty light-years away. Extra-terrestrials able to detect these signals will also gain some insight into the nature and technological state of our civilization, and if they then wished to communicate with us, they could easily do so by sending us messages on the same wavelengths they have detected. Perhaps they have already done this: some alleged alien messages received via radio will be described in Chapter 9.

In 1977 another message from Earthmen to extra-terrestrial aliens was sent out, this time on the two *Voyager* probes. These craft were headed for the outer planets of our Solar System, and have indeed sent

'The Sounds of Earth' record sent into space in 1977 aboard the Voyager *spacecraft.*

back spectacular pictures and considerable useful information from Jupiter, Saturn, Uranus and Neptune. Now they are leaving our Solar System and will fly aimlessly among the stars, though communication from Earth control should still be possible until about 2007. Early in 1990 *Voyager 2* sent back the most distant photograph of Earth ever taken: from nearly four billion miles away, the planet is a blue dot surrounded by blackness. If either *Voyager* probe is intercepted by alien intelligences, they will be able to enjoy two hours of terrestrial entertainment, for on board both probes is a twelve-inch long-playing record bearing pictures as well as sounds. The pictures show views of our Solar System, the human body, terrestrial landscapes, animals and plants, human activities and technology; the sounds include messages in eighty languages, and a selection of music, from Beethoven to Chuck Berry, Bach to Louis Armstrong, and also ethnic music like Navajo Indian chanting. The records are protected against erosion by interstellar dust particles and should last for over a billion years. But we will never know whether they have been played and enjoyed by extra-terrestrials – if they ever are, life on Earth may be long extinct by then.

Exploring outer space

Grandiose schemes for interstellar colonization by Earthmen belong in the realms of science fiction – at least until we have achieved more than landing a few men on the Moon. Plans for the immediate future are generally more realistic, but still expensive.

The radio-telescopes in use on Earth have been able vastly to increase our knowledge of the Solar System and the galaxy, but there are limitations which may be overcome by taking the telescopes off the Earth's surface. Even with Earth-based telescopes, setbacks can occur for the most mundane reasons: in late 1988 the 300-foot (90-metre) radio-telescope at the National Radio Astronomy Observatory at Green Bank, West Virginia, suddenly collapsed from metal fatigue, and was crushed beyond repair. Assistant director Dr George Seielstad explained what a blow this loss is to radio-astronomy: 'At the very moment that it collapsed we were making a survey of 100,000 radio sources in the northern sky, including some that were 10,000 million light-years away. Now much of this data will be lost for ever. The Green

Bank telescope was the only one which was surveying the entire northern sky from the equator to the North Pole, as opposed to looking at particular objects.' The only bigger steerable radio-telescope is in West Germany.[12]

Another problem facing radio-astronomy carried out from Earth's surface is the increasing radio pollution from satellites: natural radio transmissions from space are blocked by military satellites, television broadcasting satellites and even cordless telephones. A few years ago, astronomer Professor Sir Bernard Lovell complained that microwave ovens were leaking energy which interfered with incoming radio signals. The ideal solution, say radio-astronomers, would be to locate their telescopes beyond the Moon, which would shield them from interference from terrestrial satellites.[13]

The trend is now to locate telescopes in space. 1990 saw the launch of the Hubble Space Telescope into orbit 380 miles (600 kilometres) above the Earth. The eleven-ton instrument is more than a telescope, it is 'the biggest astronomical observatory ever put into space', and when initial problems are solved, it should provide astronomers with a better view of distant galaxies and quasars than is possible using Earth-based equipment. This is because the Earth's atmosphere causes distortion which makes stars appear to twinkle. Objects thirty times fainter than can be seen from Earth will now become visible; but planets like ours will still not be perceptible because bright light from their suns will make them too difficult to pick out.[14]

The Russians plan to launch a space telescope called Radioastron in 1993, at first into an orbit 50,000 miles (80,000 kilometres) high. But this distance will gradually be increased, using the rocket engines that are on board, until it is 93 million miles (150 million kilometres) away, and orbiting the Sun. It will be looking for radio traces of the Big Bang, by which the Universe is thought to have started 15 billion years ago.[15]

The Americans are also working on a major space telescope, the AXAF (Advanced X-Ray Astrophysics Facility) which will be looking for X-rays emitted by objects in space, a task that can only be undertaken from space. AXAF will be 42 feet (12.6 metres) long and 12 feet (3.6 metres) in diameter, weighing about 13 tons. It is expected to cost at least a billion dollars, and should be launched in 1995.[16]

Further exploration of the Solar System is also planned. The United

States hopes to have a permanent base on the Moon by the end of the century. One advantage for astronomers will be that telescopes sited on the airless Moon will be able to see much further than those on Earth, and even ten to thirty times further than the Hubble Space Telescope. They might be able to see as far as the edge of the Universe, 90,000 million trillion miles (145,000 million trillion kilometres) away. According to NASA engineer Dr Michael Duke:

> On an airless world, with the stars visible by day as well as by night, we shall be able to make photographic exposures of unlimited length, which enables one to see greater detail. This is impossible on earth, and cannot be done in earth orbit because every half-hour or so the telescope's view is blocked by the earth. And the moon's stable surface means that huge arrays of co-ordinated telescopes will be possible.[17]

With the experience gained from living on the Moon, scientists will then be able to work towards a manned landing on Mars. At the time of writing (late 1989), NASA's plan is to set out for Mars in 2011. The journey from Earth will take six to nine months (as compared with four days to reach the Moon from Earth) and the astronauts will need to spend at least two years there, because of the two planets' relative orbits. At present prices, the cost of the Mars mission is estimated to be £250 billion. The Russians are also planning a manned journey to Mars, and their cosmonauts have valuable experience of staying in space for long periods of time. In late 1988, two cosmonauts returned to Earth after being in the *Mir* orbiting space station for eighty minutes short of one year. But even this endurance record does not achieve the time in space which will need to be undertaken by the men travelling to Mars, and this, as well as finding a way to shield the astronauts against deadly cosmic rays, is only one of the many problems facing the Mars project.[18]

Another major problem for all spacecraft in low Earth orbit is entirely man-made, and that is the problem of rubbish in space. Items not low enough to burn up in the Earth's atmosphere will remain in space for hundreds of years, and the Earth is likely soon to have rings like the planet Saturn's. Professor Sir Bernard Lovell commented on this problem at a press conference in 1988, when he said:

There are now a million million bits of debris in orbit if you could count the pieces that are of microscopic size. They weigh a total of more than 2,000 tonnes. Russian cosmonauts routinely push bags of garbage out of their airlocks into space. This is absolutely indefensible. It is just as wicked as picnickers who throw their rubbish into the countryside. People also leave fuel tanks in space, thinking they will remain there harmlessly. What they don't realize is that fuel is still in the tank and that it will eventually explode, adding hundreds more fragments to the vast clouds of debris.

As the years pass and the problem worsens, spacecraft will be more likely to suffer a disastrous collision with a chunk, or even a small fragment, of debris.[19] It would be ironic if Man's efforts to reach the planets and beyond were jeopardized by such a seemingly trivial thing; but in fact this desire to take the easy way regardless of its effect on the surrounding environment is seen continually on Earth today. Can we imagine that Man will be any more careful once he begins to colonize outer space? It is quite realistic to envisage a scenario in which Mankind moves steadily from planet to planet, each time leaving behind a devastated wasteland.

Man also has his sights set further afield in the Solar System. Beyond Mars lies Jupiter, and the unmanned probe *Galileo* is on its way to that planet, its mission being to descend by parachute as far as possible into the hydrogen and methane atmosphere. Its 'mother craft' will also examine Jupiter's giant moons more closely than ever before.[20] This is just one of thirty-five missions planned by NASA for the immediate future. Jointly with the European Space Agency, NASA plans in 1995–6 to launch a probe that will visit Saturn and investigate the atmosphere of its largest moon, Titan. The journey will take six years, and the project has been named Cassini, after the astronomer Giovanni Domenico Cassini, who discovered four of the planet's moons in the seventeenth century.[21]

Journeys within our Solar System take several Earth years; using the present technology, journey times beyond the Solar System would rapidly escalate and very quickly become longer than a human's life-span. Instantaneous travel through time and space are the best answer, and physicists occasionally produce theories that would solve all the problems – if they could be put into practice. The latest involves 'wormholes', tunnels through space–time, which theoretically make possible huge short-cuts in space travel. Linked to this theory is the possibility of time travel whereby one could go back in time, or even into one of the parallel universes which might exist alongside our own.[22]

At present these ideas are impossible to put into practice, of course, but theorizing of this kind is not futile because of the need to be alert to the possibility of the existence of other worlds parallel to this one, invisible but perhaps somehow interconnected. It may be that even if humanity hasn't learned how to open the door, other entities have. The startling evidence to follow in later chapters may suggest that other entities do in fact use no-delay space–time travel, and may have been interacting with humanity for millennia.

4 *Have extra-terrestrials already visited Earth?*

Mankind's interest in outer space and the possibility of intelligent extra-terrestrial life would probably also be familiar to beings on other planets which had developed a similar technology to ours. If, as some astronomers suggest, there are many habitable planets in the Universe, we might expect that some of them would have by now developed intelligent life with such a high level of technology that they would long ago have solved the problems of travelling through time and space, and would have visited and/or colonized other habitable planets. Perhaps they have indeed already been to Earth, many hundreds, thousands or even millions of years ago, and left traces of their presence here. Perhaps they are the 'gods' of olden time, who came out of the sky and influenced the course of Mankind's development.

The so-called 'ancient astronauts' were a popular subject for writers in the 1970s, and many books on this theme were published, with titles like *Our Ancestors Came from Outer Space* (Maurice Chatelain), *Gods and Spacemen in the Ancient East* (W. Raymond Drake) and *Chariots of the Gods?* (Erich von Däniken). Von Däniken was perhaps the most prolific writer on this theme, but his new books are no longer best-sellers in Britain and altogether the ancient astronaut idea seems to have fallen out of favour. Did it ever have any merit?

Superficially the idea of gods visiting Earth and helping the primitive people they found here (or even, to take the interference aspect a step further, genetically engineering the development of Man from the ape) seems feasible, especially when backed up by legends of beings coming from the sky, and even by biblical quotations such as this one from *Genesis* Chapter 6: 'There were giants in the earth in those days; and also after that, when the sons of God came in unto the daughters of men, and they bare children to them, the same became mighty men which were of old, men of renown.' There are many puzzling descriptions in old writings that could be interpreted as early encounters with extra-terrestrial beings, and yet there is a danger in forcing currently popular interpretations on to these legends and thereby distorting their real meaning. Twentieth-century Man's preoccupation with outer space has resulted in the identification of these legendary gods as spacemen; yet when the same lore was studied by people for whom space travel and the possibility of extra-terrestrial life was not a viable or provocative concept, they interpreted it in an entirely different way.

Writers on ancient astronauts have also claimed to have discovered many examples of technology that could not have been within the capabilities of our terrestrial ancestors, and therefore must have been introduced from elsewhere. Such claims relegate the people of earlier ages to a state of primitive incompetence, whereas archaeological research has shown that in many respects people of early civilizations were just as skilled and inventive as people are today. The monumental masterpieces like Stonehenge, the Egyptian pyramids, Easter Island's 600 giant stone heads, and many other stone structures on a colossal scale have all been attributed to the space people, despite the fact that archaeologists are satisfied that the construction of these monuments was well within the capabilities of the peoples in whose lands they were erected. The quality of some of the stonework is undoubtedly fine, as in the ancient South American ruins, but to deny Man the ability or ingenuity to obtain these results unaided is to misinterpret the evidence of Man's achievements repeated throughout all the major civilizations.

The same applies to smaller relics which have been attributed to extra-terrestrial sources because the technology they exhibit was sup-

High-quality Inca stonework at Ollantaytambo in Peru, thought by some people to be too precise for the work of mere mortals.

posedly not available to earthmen at the time of the object's manufacture. These include the Antikythera device, a bronze instrument found in 1900 in a wrecked ship's cargo off the island of Antikythera, near Crete. Constructed around 80 B.C. and remarkable for that time, it seems to have been a miniature calendrical calculator giving past, present and future positions of the Sun and Moon, by means of at least twenty gear-wheels.

The so-called 'Baghdad batteries' were another unexpected, but not impossible, discovery. Several of these pottery jars were discovered early this century. They contained a copper cylinder, an iron rod and fragments of bitumen, and scientific tests have shown that a half-volt of electricity could have been obtained from them. The sites where these early batteries were found date to the Parthian period 2,000 years ago, and they were probably used for electro-plating small figures with gold.

There are certainly plenty of mysterious survivals from earlier ages whose purpose still eludes us. But that does not of course mean that the answer must therefore lie in outer space. One typical example is the

One of the 'Baghdad batteries', an early scientific achievement by terrestrial man.

Nazca lines in Peru. On a bare plateau covering 200 square miles (320 square kilometres) huge patterns up to ⅝ mile (1 kilometre) in length were marked out in the earth, the designs being only discernible from the air as geometrical shapes and animals, nearly 800 of the latter. Erich von Däniken saw the Nazca lines as possible landing strips for the aircraft of the gods. As he theorized in *Return to the Stars*, 'At some time in the past unknown intelligences landed on the uninhabited plain near the present-day town of Nazca and built an improvised airfield for their spacecraft which were to operate in the vicinity of the earth. They laid down two runways on the ideal terrain.'[1] The native peoples, in an attempt to get the 'gods' to return, later made new lines on the plain, with figures of birds to symbolize flight.

Other researchers have denied the feasibility of von Däniken's theory, although no one still knows for certain why the figures were laid out. They cannot even be accurately dated, though they are thought to

A section of the Nazca lines in Peru, showing a series of parallel lines and a monkey figure. Erich von Däniken saw landing strips for the gods' spacecraft at Nazca.

have been made between 1 and 1000 A.D. Markers of astronomical events is one popular explanation for the lines, but research has not confirmed this to any great extent. Other suggestions include routes for ceremonial events, or symbols used in shamanic rituals; but whatever the true answer, it is not likely to include the landing and take-off of extra-terrestrial spacecraft, because a developed technology would be unlikely to need a runway to land on or take off from, and the sandy soil is a quite unsuitable surface for a heavy vehicle to land upon.[2]

Other evidence put forward by proponents of the 'ancient astronauts' theory takes the form of representations of possible spacemen. There are a number of examples, including the rock paintings in the Tassili N'Ajjer mountains of the African Sahara. Erich von Däniken found a Martian in a spacesuit with a heavy helmet, and astronauts with antennae on their helmets (interpreted by others as women carrying baskets on their heads). The frescoes date back to between 8000 and 6000 B.C., and tools and weapons have also been found which originated with the civilization that made the paintings.

It is equally unlikely that a tomb lid in a pyramid in the ancient Maya city of Palenque, in Mexico, bears a stone carving of a rocket pilot, although Erich von Däniken interpreted it that way. He saw a 'crouching being . . . manipulating a number of undefinable controls and . . . [with] the heel of his left foot on a kind of pedal.' He wears a space helmet 'with the usual indentations and tubes, and something like antennae on top.'[3] Seen through the eyes of a believer in ancient astronauts, the figure can indeed be seen as a 'space traveller' (as von Däniken calls him), but in fact the slab is a tribute to a Maya king who died in the late seventh century A.D., and follows a purely religious theme.

Some early writings have also suggested 'ancient astronaut' themes, and especially extra-terrestrial spacecraft, like the celestial chariots described in various Sanskrit writings of 8,000 or so years ago, and, more recently, the wheeled aerial craft seen by Ezekiel the priest and described in the Bible. Ezekiel, who lived 2,500 years ago, began his account of his experience, 'And I looked, and, behold, a whirlwind came out of the north, a great cloud, and a fire infolding itself, and a brightness was about it, and out of the midst thereof as the colour of amber, out of the midst of the fire.' He went on to describe 'four living creatures', composite animals and human beings interpreted as cherubim, and a wheeled craft of some kind. NASA engineer Josef Blumrich investigated Ezekiel's vision after reading about it in von Däniken's *Chariots of the Gods?*, and initially was sceptical that Ezekiel was describing a spacecraft, as postulated by von Däniken. But as he investigated further, so he too saw the vision as an extra-terrestrial craft, and wrote up his interpretation in a book, *The Spaceships of Ezekiel*.[4] In fact it is highly unlikely that the extra-terrestrial of 2,500 years ago would have been building spacecraft very similar to those envisaged today by earth scientists; and Ezekiel himself says that he saw 'visions of God', thus clearly signifying that the sights he saw were symbolic rather than real images.

The case of the Dogon, a West African tribe, and their knowledge of the Sirius star system is also open to more than one interpretation. The information possessed by the Dogon was researched over seven years by Robert K. G. Temple, who then wrote up his results in his book *The Sirius Mystery*.[5] The Dogon mythology told of the landing of amphibious creatures, the Nommos, several thousand years ago in an ark. The Nommos came from Sirius and proceeded to pass on astronomical information about their home to the Dogon, including the fact that the star Sirius has a companion star which we cannot see, and which revolves in an elliptical orbit around Sirius. This is true; Sirius B does exist in an orbit around Sirius. They also told of a third star in the system, but astronomers have not yet located this star.

The case presented by Temple is very detailed and complex, and has been hailed by some people as conclusive proof that ancient astronauts have indeed visited the Earth. However, others are more cautious, and have been able to counter Temple's arguments. For example, it is likely that the Dogon could have got their astronomical knowledge elsewhere than from the Nommos. They are not totally isolated from the outside world, and there have been European and Islamic schools in the area this century. So the special knowledge of the Sirius system, allegedly known by Dogon priests for centuries but not discovered by Western astronomers until 1862, could have been learned by the Dogon and incorporated in their mythology much more recently. There is also other astronomical evidence which tells against the idea that intelligent beings came to Earth from Sirius.[6]

Virtually all the individual bricks from which has been built the 'ancient astronauts' hypothesis have been shown to be made of straw, once they were thoroughly investigated by impartial researchers. Other 'evidence' includes the Ark of the Covenant (an electric condenser?), Jesus as a spaceman, the giant stone heads sculpted by the Olmec (astronauts?), the Cabrera stones ('ancient' stones engraved with modern scenes, and of dubious antiquity), the Tunguska explosion (a crashing spaceship?), and the Chaldean legend of Oannes (a fish-bodied deity who founded civilization): all these mysteries and others are exposed in Ronald Story's two books *The Space-Gods Revealed* and *Guardians of the Universe?*

Modern mysteries in ancient stones

But there does remain one category of enigma which has not proved easy to explain away, and which could possibly be interpreted as evidence for extra-terrestrial visitations in the distant past. This evidence comprises cases where some apparently modern object has been found inexplicably sealed inside a lump of rock, and also the finding of seemingly human footprints fossilized in rock millions of years old. Sometimes the footprints are said to be distorted animal tracks, sometimes they are found to have been carved by native peoples; but sometimes there is no straightforward explanation, especially when the prints are found after some overlying rock has been removed, in which case they could not be recent carvings. However, footprints found in rock do not necessarily always predate the earliest humans known to have lived on Earth, for some types of rock have formed relatively recently.

When objects are found inside rocks, several interpretations are possible. The finds could indicate that the rocks are not as old as has been believed: this is a view held by the creationists to support their literal interpretation of the Bible. Alternatively, the finds could indicate that Man has been around, and with a viable technology, for much longer than is generally realized. A third interpretation is that extra-terrestrials brought their technology to Earth, long before Man evolved. But it is always possible, of course, that some or all of these discoveries are hoaxes. Here are a few examples of these enigmatic finds.

In Dinosaur Valley State Park, Paluxy River, Glen Rose, Texas, USA, dinosaur tracks were believed to have been found interspersed with human footprints, and some of them are visible in the foreground of this photograph.

In the late 1780s, in a quarry near Aix-en-Provence in France, coins, hammer handles, other tool fragments and broken boards were found, all 'changed into agate'. These finds were 50 feet (15 metres) deep, 'and covered with eleven beds of compact limestone'.[7] In November 1830 a block of marble was quarried from a depth of 60–70 feet (18–21 metres) in a quarry near Norristown, Pennsylvania, and when slabs were cut off the block, two raised characters, like the letters I and U, were found in an oblong indentation 1½ inches (3.75 centimetres) long and ⅝ inch (1.56 centimetres) wide in the marble.[8] A nail was found in a block of stone in Kingoodie quarry near Dundee, Scotland, possibly in 1844 or a short while earlier, and the circumstances of the discovery are given in this contemporary report:

> The stone in Kingoodie quarry consists of alternate layers of hard stone and a soft clayey substance called '*till*'; the courses of stone varying from six inches [15 centimetres] to upwards of six feet [1.8 metres] in thickness. The particular block in which the nail was found, was nine inches [22.5 centimetres] thick, and in proceeding to clear the rough block for dressing, the point of the nail was found projecting about half an inch [1.25 centimetres] (quite eaten with rust) into the '*till*', the rest of the nail laying along the surface of the stone to within an inch [2.5 centimetres] of the head, which went right down into the body of the stone. The nail was not discovered while the stone remained in the quarry, but when the rough block (measuring two feet [0.6 metres] in length, one [0.3 metres] in breadth, and nine inches [22.5 centimetres] in thickness) was being cleared of the superficial '*till*'. There is no evidence beyond the condition of the stone to prove what part of the quarry this block may have come from . . . It is observed that the rough block in which the nail was found must have been turned over and handled at least four or five times in its journey to Inchyra, at which place it was put before masons for working, and where the nail was discovered.[9]

Also in 1844, a gold thread was found embedded in stone 8 feet (2.4 metres) deep by workmen quarrying rock close to the River Tweed in southern Scotland.[10] A beautiful work of art, a bell-shaped vessel, was found among fragments after a rock was dynamited at Dorchester, Massachusetts, in 1852. The vessel was 4½ inches (11.25 centimetres) high and 6½ inches (16.25 centimetres) across at the base, and appeared to be made of a silver alloy:

Six figures of a flower, or bouquet, beautifully inlaid with pure silver, and around the lower part of the vessel a vine, or wreath, inlaid also with silver. The chasing, carving and inlaying are exquisitely done by the art of some cunning workman. This curious and unknown vessel was blown out of the solid pudding stone, 15 feet [4.5 metres] below the surface . . .[11]

Some years before 1883, when details of the find were published, a Colorado man obtained coal from a drift mine known as the Marshal coal bed, the coal being taken from a point 300 feet (90 metres) below the surface. The report continues:

Upon my friend's return home he placed some large chunks of the coal in the stove, but upon its not burning well, he broke them and in the midst of one, imbedded in a hollow place, but completely surrounded by the coal, the thimble was found. These coal beds are classed by Prof. Hayden as lignitic and lying between the Tertiary and the Cretaceous. Much of the coal is 'fresh', some of it too 'green' to burn well. My informant says the chunk in which the thimble was found 'showed the grain of the wood'. For some time he kept it, but it is now lost. The thimble was full of coal and sand and retained its shape well.

The thimble was of iron and moulded, and began to crumble away with much handling.[12] In 1891, in Illinois, Mrs S. W. Culp was shovelling coal when a lump broke open and she saw a gold chain lying in a cavity in the coal.[13] Mrs Myrna Burdick found a spoon among ash from burnt coal in 1937, at her Pennsylvania home. The ashes had not been disturbed after a large piece of coal was burned, but when they fell apart, the spoon was noticed among them.[14]

If all these discoveries really took place as described by the contemporary reporters, these hidden objects are likely to be only a small proportion of a much greater number which still remain locked away inside solid rock. A similar mystery which also defies explanation is the finding of toads and frogs still alive in cavities inside rocks (see our earlier book *Modern Mysteries of Britain*). It is impossible to say for sure what these mysteries signify, but we cannot rule out the possibility that the finds of manufactured objects are telling us that Man and technology have been around for much longer than anyone has realized. However, we would be cautious about attributing the finds to the space people. Would they really have been here on Earth, using spoons and thimbles, in the age of the dinosaurs at the time the coal beds were laid down? If extra-terrestrial entities were on Earth aeons of time ago, they were highly advanced technologically, and it is almost certain that their technology would not be recognizable to us; any artefacts that survived and were preserved in rock would therefore probably be inexplicable to us, as well as being constructed from unidentifiable, or certainly non-terrestrial, substances, but so far as we know, no such objects have been found.

Nor is it likely that any of the more recent historical mysteries, such as are described earlier in this chapter, has any connection with visitors from outer space, for reasons already given. In fact, there is *no* conclusive evidence that the Earth was visited by anyone from other planets in the distant past. If they did come here, they have not left us a message that we have found or can read.

Would they have been likely to leave us any message? If their thinking was along the same lines as ours, they might have wanted to leave a sign of their existence to any future occupants of Earth (assuming there was no human life on Earth when they came), or if they were themselves responsible for the development of intelligent Mankind on Earth, they might have wanted to leave a message for us that they had been on Earth at our birth.

Where would they leave such a sign? Edward Ashpole points out in his book *The Search for Extra-Terrestrial Intelligence*[15] that any mark left on Earth would have eroded away within a few thousand years, and so intelligent beings might have left their mark on the erosion-free Moon, which mark could remain unchanged for a billion years or more. But so far, no such sign has been found on the Moon. Perhaps when Man begins to explore it more thoroughly, and to take up long-term residence there, something may be found. There is still the possibility that ancient astronauts got there first.

5 Are extra-terrestrials visiting us now?

The evidence for extra-terrestrial visitation before the present century appears to be so ambiguous as to be virtually worthless. Far more intriguing are the thousands of reports of possible extra-terrestrial visitors this century – those who travel in the so-called 'flying saucers' or 'unidentified flying objects' (UFOs). The phenomenon of alien aerial craft came into the consciousness of Mankind in 1947, following Kenneth Arnold's famous sighting on 24 June of 'a formation of very bright objects' travelling fast in the Mount Baker/Mount Rainier area of Washington state. Arnold was alone in a small plane, searching the high plateau of Mount Rainier for a crashed transport plane. He was fascinated by the 'aircraft' he saw which, as he put it to newsmen later, 'flew like a saucer would if you skipped it across the water'.[1] Thus was born the term 'flying saucer', and from that time sighting reports have been widely published in the media, though in the 1960s the preferred term became the more scientific-sounding 'unidentified flying objects' (UFOs).

From the earliest days the objects (or lights, as many sightings are of nothing more substantial than lights seen in a night sky) were interpreted as craft from outer space, probably carrying alien beings. This was a familiar concept, having for several decades been used by science fiction, and it seemed a logical assumption that non-terrestrial craft must therefore be extra-terrestrial, though this reasoning is in fact faulty. Many people assume that 'UFO' means 'extra-terrestrial spacecraft', but it does not naturally follow that UFOs are craft, nor that because they are seen in the sky, they come from 'somewhere out there'. Some researchers would argue that no UFOs are solid objects. Many alternative explanations exist, and people clearly do misidentify mundane events: lights in the night sky could be attached to aircraft (terrestrial), they could be stars and planets, or sometimes the Moon (which when seen veiled in cloud can look very mysterious). Other stimuli mistaken for spacecraft include ball lightning and other rare electro-magnetic phenomena, comets and meteors, assorted rare meteorological events, balloons (weather and toy) and kites, birds (especially when catching the sunlight or illuminated by street lamps at night) . . . It is clear that there are many mundane explanations which should be considered before more esoteric ideas are put forward. And it should not be forgotten that some sightings or reports could be hoaxes, as could some of the UFO photographs. The truth is that although there are on record literally thousands of UFO sighting reports, very few of them cannot be satisfactorily explained in terms of our known terrestrial science, if experienced researchers devote sufficient time to them. Yet some truly mysterious cases do survive close scrutiny, and it is not possible for anyone seriously to deny the existence of the UFO phenomenon.

Extra-terrestrial hypothesis

However, there is at present no incontrovertible evidence that any UFOs come from outer space. (Their pilots and crew sometimes say so – see Chapters 6–13 – but can they be believed?) Nevertheless, the ETH (extra-terrestrial hypothesis) has taken firm root, so much so that many people do not question it. During recent years polls have been taken to discover the extent of people's beliefs in occult and esoteric subjects, and one taken in 1976 revealed that belief in the ETH is by no means confined to the less intelligent members of society. In the USA a poll of MENSA members (i.e. people with a high IQ) showed that two-thirds of them believed that UFOs are 'spaceships from another planet', and these believers said they would 'accept an offer to board a UFO if its occupants were not visibly hostile'. Fifty-one per cent believed that UFOs are 'carrying passengers who are studying our behaviour', and sixteen per cent of respondents said that they had themselves seen a UFO.[2]

That belief in the existence of extra-terrestrial life is firmly established was demonstrated as long ago as 1938, when the famous Orson Welles radio broadcast of H. G. Wells's *War of the Worlds* was made.

More than a million of the twelve million American listeners seem to have missed the pre-broadcast warnings that the programme was a dramatization, and believed the 'news bulletins' that Martians had landed at Grovers Mill in New Jersey. Thousands rushed out into the streets, and huge traffic jams built up as they fled. Farmers around Grovers Mill grabbed their guns, and one shot the water tower, thinking it was the enemy spacecraft.[3]

In the fifty years since then, American folk have continued to show their support for the ETH in a creative way: some of them have built model spacecraft to stand outside their houses, others have constructed UFO landing ports. Some perform rituals to summon UFOs to land, while a few try to raise the spiritual awareness of Mankind in preparation for the landing of spacemen on Earth.[4] One of the latest ventures took place at Elmwood in Wisconsin, where in 1988 Tom Weber founded a group called UFO Site Center Corp with the aim of building a safe landing site for spaceships. He said, 'There have been numerous sightings of unidentified flying objects, particularly spaceships. We believe these spaceships are manned by an intelligence that has been watching us for a long time. They are simply waiting for us to take the next step and give some kind of invitation to them.'[5]

There is clearly widespread belief in the validity of the extra-terrestrial hypothesis, so what are the arguments put forward by those UFO researchers who care to defend it? The major arguments are:

1. The large number of UFO sightings, and the reliable nature of the witnesses.
2. The frequent confirmation of UFO activity in the form of ground traces, photographs, electro-magnetic interference with vehicles, radar images, etc.
3. Government interest and involvement: high-level scientific interest, claims of crashed UFOs and bodies retrieved and studied.
4. Meetings between UFO witnesses and UFO occupants, the latter sometimes claiming to come from outer space (see Chapters 6–13).
5. The large number of abduction cases, with details corresponding from case to case.
6. The possibility that civilizations of intelligent beings exist else-

where in our and other galaxies, and the probability that they would want to explore space and visit inhabited planets.

All these arguments can easily be met by equally valid counter-arguments:

1. Many UFO reports have been shown to describe natural events, not spacecraft, and witnesses are often unreliable for various reasons.
2. All the apparent confirmations could have other causes than the presence of alien spacecraft, e.g. hoax, natural phenomena.
3. There is no solid evidence of government involvement, only rumour.
4. Witnesses who claim to meet UFO occupants may be lying or hallucinating. Or they may be seeing another kind of entity, which only claims to be from outer space, whereas its true origins lie elsewhere.
5. Abduction cases are very suspect, for the details are often revealed under hypnosis, an unreliable method of obtaining factual information.
6. As already shown earlier, there is at present no scientific evidence that there is anyone else 'out there'; and even if there were intelligent life somewhere in space, we should not assume that they have the same motivations as humanity.

Let us assume that some UFO sightings are indeed of alien spacecraft. If we look more closely at what is apparently happening, based on witness reports, further illogicalities immediately become apparent. There have been literally thousands of UFO reports in the forty-odd years since Kenneth Arnold's sighting, of which a large number are very probably attributable to mundane stimuli. If those for which no explanation can be found are indeed alien spacecraft, we must ask why this planet has had so many visits from alien entities. If Man were to visit even the relatively close Moon as frequently, it would divert such a huge proportion of the Earth's resources into that one project that it would become an intolerable burden upon humanity. The craft vary tremendously in shape and size, so it is not always the same craft that is being seen. Does this mean that the visitors come from many different planets? If so, why are all the extra-terrestrials so interested in Earth, which is after all a small planet revolving around a small sun near the outer perimeter of a moderate-sized galaxy in a universe of billions of galaxies.

They have been visiting us for at least forty years, but their intentions are still incomprehensible. Much of the activity reported of landed aliens seems either purposeless or inexplicable, except for those who appear to be gathering samples (plants and animals – and humans? See Chapter 7 for further information). So many craft over such a long period of time indicates that there must be vast back-up facilities somewhere, and vast factories making more and more spacecraft – for what reason?

As soon as we start to examine the 'evidence', using terrestrial logic, many difficult questions arise which need answering, and all attempts at answers must involve a large degree of speculation or even fantasy: the aliens do not have any technical or financial constraints on production, as humanity does; or possibly the craft and even the aliens themselves are simply thought-projections, not 'real' or physical. They appear in whatever guise is most acceptable to us – that's why their craft designs often echo terrestrial ideas of what a spacecraft should look like. Because their motivations are so different from those of humankind, their actions will always be quite inexplicable. Perhaps Mankind is their experiment, and is kept under observation by frequent visits. There are plenty more ideas along these speculative lines, and they are only justified as attempts to come to terms with the amazing variety of material obtained from the UFO reports of the last forty years.

Puzzling UFO reports

It seems clear that there is little to be gained from theorizing – the UFO mystery is not likely to be solved that way. Our present intention in this book is to ascertain whether UFO sightings provide any evidence at all that there are alien spacecraft visiting Earth. We must stress again that a large percentage of the so-called 'UFOs' are explicable in terrestrial terms, most being misidentifications of normal objects or events. However, there are some intriguing cases in the records which do not

One of the McMinnville UFO photographs taken in 1950.

easily lend themselves to mundane explanations. If the witness really did see what was reported, and was not hoaxing or suffering from imperfect eyesight or hallucinations, then the following cases seem to provide evidence that 'someone else' is in our skies.

A classic sighting of over forty years ago which has survived critical analyses took place at McMinnville, Oregon, on 11 May 1950. The witnesses were Mr and Mrs Paul Trent, who saw a flying disc from their small farm, and took two photographs of it. These photographs have stood up to the close scrutiny of photographic analysis using computers, with no signs of hoaxing to be found. Also arguing against a hoax is the fact that the couple did not rush to get the film developed and they only showed the pictures to their family. It was a while before the photographs found their way to the media, and this was not instigated by the Trents. It has been calculated that the object was about half a mile (1 kilometre) away, and so would have been 100 feet (30 metres) in diameter and 13 feet (4 metres) thick. It appears to be a silvery,

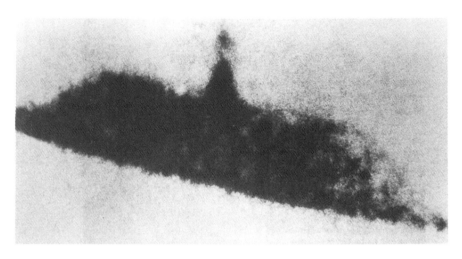

metallic, disc-shaped flying craft of artificial construction – but forty years on we do not know what it was, where it came from or where it went.[6]

Another important photographic case dates back to 1958. The Trindade Island photographs have not yet been conclusively proved not to show an alien craft, despite serious attempts to explain them in natural terms. Trindade Island is a small rocky island in the South Atlantic Ocean used as a US and Brazilian military base during the Second World War. Late in 1957 a Brazilian Navy boat went there to build a meteorological station, and was preparing to leave on 16 January 1958 when the UFO was seen. It moved quickly towards the island, hovered over one of the rocky peaks, then went briefly behind it before being seen to move away to sea.

Photographer and former newspaperman Almiro Barauna was on deck with a camera when the UFO appeared and he managed to get four photographs of it before it was finally lost to view. There were other witnesses on deck, about a hundred sailors. When questioned later, Barauna gave more details of the object: 'It showed a dark grey colour, appearing to be surrounded – mostly in the area ahead of it – by a kind of condensation of a greenish, phosphorescent vapour (or mist).' He also said that it appeared to be metallic, and 'was obviously a solid object'. As for its mode of flying, 'It showed an undulatory movement as it flew across the sky, like the flight of a bat. And when it came back, it changed speed abruptly, with no transition, in a jump.' After the sighting,

Barauna processed his film at once, but having no printing paper on board he could not make any prints from the negatives until he got back to Brazil. Afterwards, Navy officers calculated that the object was about 120 feet (36 metres) in diameter and 24 feet (7 metres) high, flying at about 600 miles (965 kilometres) per hour.[7]

In the thirty-odd years since the events described, attempts have been made to discredit the photographs and prove them to be hoaxes, but without success. The latest attempt to explain them was made by researcher Steuart Campbell in 1989. He concluded that they were not likely to be hoaxes: fakes usually consist of a single picture, not four, and when joined up Barauna's pictures coincide and show the object moving out to sea, as described by witnesses, who incidentally included senior Navy officers. Campbell's hypothesis is that the pictures show a mirage of the planet Jupiter, more precisely 'a double-merged magnified mirage of Jupiter', and although he admits that his hypothesis may seem unlikely, he feels it is the only feasible explanation. Other researchers do not agree. Campbell has ruled out an alien craft of any kind, because he simply cannot believe in them, but in view of other mysterious and inexplicable sightings, we feel it is hasty to discount this possibility.[8]

Of the many reports of landed UFOs, one that has long fascinated researchers took place on 19 May 1967. The unfortunate witness was Stephen Michalak, who went to Falcon Lake in Manitoba, Canada, to do some amateur prospecting. While he was sitting eating his lunch, his attention was drawn to two UFOs coming down out of the sky. They had domes on top, and glowed red. One flew off; the other hovered close to the ground about 150 feet (45 metres) away from Michalak. He

A UFO photographed in 1958 by Almiro Barauna, one in a series he took of an unidentified object flying over Trindade Island.

Stephen Michalak's drawing of the UFO he touched when it landed in Canada in 1967.

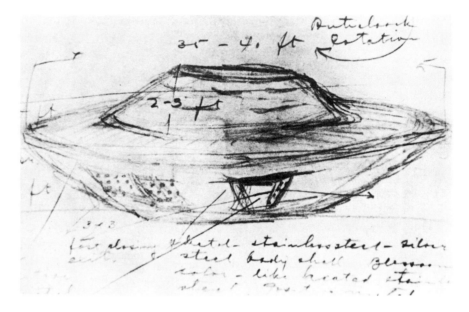

Stephen Michalak's drawing of the UFO he touched when it landed in Canada in 1967.

watched it for about half an hour, during which time he made a drawing of it. A bright purple light was being emitted from slits near the object's top, and Michalak also noted that its colour changed from red to steel-grey, that it was giving off waves of warm air, and a sulphurous smell.

When he saw a door open in the side, he went closer. He heard human-like voices, and saw lights 'like on a computer panel'. He called out to the occupants (whom he could not see), using several languages. At this time he still thought the craft was a terrestrial one, perhaps an American experiment of some kind. When he went right up to it and touched it with his gloved hand, the glove melted from the heat. Then the door closed (he had still seen no occupants), the object rose and rotated, then gave off hot exhaust fumes which hit Michalak on his chest and set his clothes on fire. He pulled off his burning shirt as he watched the craft move away and out of sight. He felt very ill, with a headache and nausea, but managed to walk back to civilization where he sought medical help.

Several theories have been aired, but scientific investigation has not come up with a firm explanation. Researchers do not believe Michalak hoaxed the case: he still has the scars from his burns. But did he really see an alien spacecraft, or a terrestrial experiment of some kind as he

64

thought at the time, or did he have a complex physical/psychological experience, in which close contact with some natural magnetic phenomenon produced in the rocks at the site affected him mentally and resulted in hallucinations? This would not explain the burns, but the 'UFO' may have been an outburst of energy from the phenomenon, lacking form but perceived by Michalak as a craft. Far-fetched? Maybe. It all depends on whether you consider any of the other explanations more or less unlikely.[9]

It is always vital, if a UFO landing is to be investigated thoroughly, that samples of soil and vegetation from the landing site are gathered promptly and efficiently and submitted for professional analysis. In the Michalak case, there were suggestions that rock samples taken from the site had proved to be radioactive, but as there were deficiencies in the investigation, the results are ambiguous. A possible UFO landing in France in 1981 benefited from the fact that at that time there was an official UFO study group, known as GEPAN and set up in 1977 to investigate and analyse reports received from the gendarmerie, the military and scientific civilian sources. The landing took place in Trans-en-Provence, and there was a single witness, Monsieur Collini, in whose garden the ovoid object landed. He had been attracted by the low whistling noise it made as it approached, and it continued to make this noise as, after only a minute on the ground, it rose again and moved away. Monsieur Collini went to the landing site and saw a crown-shaped impression on the ground as well as circular marks. Next day the gendarmerie came and took samples of soil and vegetation which were analysed by GEPAN. They found that heating of the ground to 300–600°C had occurred, and small quantities of such materials as phosphate and zinc had probably been distributed. It was also found that there had been a thirty to fifty per cent reduction in the chlorophyll in the plant leaves, possibly caused by an electro-magnetic field.[10]

Clearly something had landed at Trans-en-Provence, but what? Was it a UFO, or again, as suggested in the Michalak case, some electro-magnetic phenomenon which affected Monsieur Collini and caused him to hallucinate a solid craft? But could such a phenomenon have left landing traces on the ground? It seems unlikely.

These four UFO reports have stood the test of time, and despite repeated analysis have failed to be proved hoaxes or caused by misidentification of mundane stimuli. But they nevertheless cannot be used as conclusive evidence that UFOs are extra-terrestrial spacecraft. Although it is tempting to jump to that conclusion, it is too long a jump from 'strange craft in sky' to 'visitors from outer space'. There is, for example, no evidence at all that these craft ever travelled out of Earth's atmosphere; if really alien, they may still have come from close at hand – an alien base on Earth? Or from a parallel universe or another dimension?

UFOs in space

There *are*, however, reports of unidentified objects being seen in space, and these stand more chance of being extra-terrestrial in origin. As they have not been seen at close quarters, though, it is not really possible to conclude that any of them are solid, manufactured objects, and they might all be natural phenomena of some kind. There are many possible identifications for the numerous reports of dark spots seen moving across the Sun, both astronomical (meteors, asteroids, etc) and non-astronomical (birds, insects, hail, aircraft, satellites, etc). One recent case dates from 14 January 1983, when the witness, based in Winnipeg, Manitoba, Canada, saw 'a perfectly round black orb crossing the sun . . . [it] lasted three seconds. On a projected solar disk with a diameter of 18 centimetres [7.2 inches], the object had a diameter of one-half centimetre [0.2 inch].' It was moving too fast for a planet, too slow for a meteor.[11] Similarly, dark objects have been seen crossing the Moon's surface, and small natural Earth satellites, probably of a temporary nature, may be responsible for some of these sightings.[12] Positive identification is difficult, as also in the case of a 'faint star-like object' seen on 8 September 1956 by a man observing Mars through a telescope from Las Cruces, New Mexico. As he reported his sighting:

> The sky was partly cloudy but the planet could usually be observed in full brilliance for several minutes at a time between clouds. At this time a faint star-like object of about eleventh magnitude was observed to pass through the field of vision in a direction opposite to the apparent drift of Mars and about one minute of arc below the lower limb of the planet. This corresponds to a true position of about one

minute of arc above Mars and travelling east . . . The telescope was moved so as to follow the object which once again was observed to transit the field in about ten seconds . . . As far as could be ascertained the object was of a yellowish colour.[13]

Although most of the sightings of mystery objects in space are made through telescopes, sometimes particularly bright objects are seen with the naked eye, as happened on 9 April 1983. The witness of this occurrence was on the *Dorsetshire* in the North Atlantic Ocean when he saw a bright white object in the sky.

It was moving rapidly southwards across the sky, leaving a bright trail behind it, like an afterglow. Also trailing astern of the object was a light trail of sparks (possibly large solid particles). The object disappeared behind clouds, bearing about 170° (T) at an elevation of approximately 35°, and lighting the edges of the clouds. The time taken for the passage was around twenty seconds. It was obviously a very large object, judging from its apparent size as seen from sea level. The impression given was that of an object within the atmosphere, easily showing around a one-penny piece held at arm's length.

This could have been a fireball, but the flight direction of north to south is unusual, and a fireball would be expected to pass over in a much shorter time than twenty seconds.[14]

It might be expected that astronauts in flight would be particularly well placed to observe any UFOs flitting around close to the Earth, and indeed from time to time rumours of such sightings have come into existence. However, after careful investigation it has been found that all the alleged astronaut sightings of UFOs so far can be ascribed to hoaxes or misidentifications of known objects. The most recent case dates from March 1989 when it was claimed that one of the astronauts on the American space shuttle *Discovery* made a transmission which included the words: 'Houston, we still have the alien spacecraft under observance.' Representatives of MUFON (Mutual UFO Network) in the States investigated the recording immediately, and were told by NASA that, 'We believe that this is a fictitious event and is a hoax perpetrated by a rogue radio operator or an unlicensed person using radio equipment and broadcasting on the repeater frequency that some ham groups use to relay NASA transmissions.' When the *Discovery* pilot, John Blaha, was interviewed by MUFON investigator Bob Oechsler on his UFO radio programme broadcast from a Baltimore, Maryland, radio station, he was asked about the transmission and emphatically denied having made any comment about alien spacecraft. Likewise the *Discovery* commander, Michael L. Coats, who wrote in a letter to Philip Mantle, MUFON Representative in England:

> The reason you will *never* hear an *actual* tape recording of any of us on *Discovery* discussing 'aliens' is because we *never* did. The stories are amusing, but pure fiction. They are manufactured to sell magazines and newspapers . . . If we *did* see any aliens, the whole world would hear about it *immediately*. We are just as curious about the possibility of other life as *anyone*, so why would we try to be secretive?

Voiceprint analysis was also undertaken by MUFON, and there was found to be insufficient evidence to link the voice on the 'alien spacecraft' tape to that of any of the five members of the *Discovery* crew.[15]

The foregoing cases, and especially the possible fireball, demonstrate the difficulty of drawing a dividing line between natural phenomena and so-called UFOs (i.e. solid craft). The percentage of UFOs that definitely appear to be manufactured craft of a design not known on Earth is really quite small, but so long as such cases do exist (for example the McMinnville case described earlier, where the witnesses' word cannot be doubted as the object was also captured on film) the theory that such craft are interplanetary can be retained. Despite the fact that this theory has been around for more than forty years, during which time ufologists have collected thousands of sighting reports, the accumulation of all these data has not resulted in any progress whatsoever towards discovering once and for all whether there are 'men from Mars' (or elsewhere) piloting the craft. A *belief* that this is so is a long way from *scientific proof*.

Perhaps in frustration at the lack of progress, some ufologists have turned their attention towards the government. Their reasoning is that since authority is in control, it must know what is going on; and from this has arisen the firm conviction among some ufologists that the American military possess alien craft which have crash-landed on Earth, together with dead, or perhaps even living, crew members. Naturally the authorities deny this, which in turn confirms the paranoid suspicions of the crashed-UFO enthusiasts that there is a massive cover-up of the truth. The populace would panic if presented with conclusive proof of the existence of space aliens, says one theory for the reason behind the cover-up. Despite intensive and diligent research on the part of some UFO researchers who have taken a special interest in the theme of crashed UFOs, yet again conclusive proof is still lacking. In fact this is an important aspect of UFO research, for the existence of a crashed non-terrestrial spacecraft would demonstrate conclusively that someone is visiting us from somewhere not on this Earth. There are on record at least twenty-eight reports of UFOs crash-landing between the years 1942 and 1978, and from these crashes a total of 102 bodies are said to have been recovered.[16] There are crash locations worldwide – South Africa, England, the Sahara Desert, Bolivia – but they are mainly in the United States of America, in the deserts of New Mexico and Arizona. Why UFOs should happen to crash so often in these deserted areas is puzzling.

The most famous of these alleged UFO crashes is the one that took place near Roswell, New Mexico, on 2 July 1947. The location was a very isolated ranch – the nearest town being 30 miles (50 kilometres) away. An early press release only a few days after the crash claimed that a flying disc had been recovered, but on the same day this statement was made, it was retracted and the object was said to be nothing more than a weather balloon. However, great efforts were made by the military to retrieve all the pieces and to keep details of it secret, and this has naturally led to continuing belief that it really was an alien spacecraft that crashed in the desert.[17] Researcher Ron Schaffner has suggested that a more plausible explanation is that it was debris from a crashed V-2 rocket. The Army Ordnance Department fired 103 missiles during the ten years from 1945 to 1955, including 67 German V-2 rockets at White Sands, and several of these went astray. Schaffner believes that the military wanted to keep their activities secret and so devised a highly successful misinformation plan, involving UFO reports that were hushed up, this naturally followed by cries of 'Conspiracy!' As Schaffner commented, 'Actually it is a cover-up twice over, much like a double agent. It is an excellent ploy and has worked for nearly forty years.'[18]

Of course Schaffner's theory is controversial, many researchers resolutely holding to their belief that the first press release told the true story, and that somewhere the government has at least one crashed UFO hidden away. If this were the case, and if it were also true that at least twenty-eight UFOs have crashed, then surely someone, somewhere, would have leaked the information by now. Many people must have been involved in the retrieval operations and the subsequent research work, and despite their oaths of secrecy someone would have revealed something. According to one researcher, Leonard H. Stringfield, this has indeed happened, and he has himself located and interviewed many people who claim to have taken part in such work. Stringfield has amassed a large file of data obtained from such people, and has published several reports, yet somehow the truth remains just beyond our grasp. Witnesses use pseudonyms, ostensibly for fear of official retribution, and confirmation of their stories does not seem easily obtainable; for us the crash/retrieval tales remain no more than rumours.

A more recent crash than Roswell, which has received wide publicity, happened at Kecksburg, Pennsylvania, on 9 December 1965.

The object was first seen streaking across the sky, with thousands from Michigan to New York witnessing a brilliant ball of fire which left a smoke trail, visible for about twenty minutes after it passed.

Many, including pilots who observed it, thought it was an aircraft which was on fire. Reports of debris from the object were made in many states, and an Ohio fire department was called to extinguish ten small fires in an area where witnesses said they saw flaming fragments falling from the sky.

Shock waves were reported by pilots, and a seismograph near Detroit recorded a shock . . .

A young fireman was one of those called in to search for the object, thought to be an aircraft. In 1989, he remembered:

It was getting semi-dusk and we had flashlights. We were taken in the back of a truck and dropped off and told to go 'this way', which we did. I was not on the initial contact team. Another team found the object.

It was definitely, unequivocally, positively, absolutely no aircraft, plane, helicopter or rocket, at least not to my knowledge. It was in an area that was part field and part woods and we went down to investigate. We found the object had crashed at a thirty to forty degree angle, and had broken off numerous tree branches in its impact path. My initial reaction was 'This is no airplane.' I observed no shrapnel, no breaking up of the fuselage. It was one solid piece, no doors, no windows . . . I've been a machinist for twenty-four years and I've worked with a tremendous amount of different metals, and I have never seen any type of metal that looked even close to that.

He also said that the visible debris was 8–10 feet (2.5–3 metres) long, 6–7 feet (about 2 metres) across, shaped like an acorn lying on its side. It was not cracked or dented, and did not give off smoke or steam. He responded negatively to the suggestion that it was a meteorite:

It had writing on it, not like your average writing, but more like ancient Egyptian hieroglyphics. It had sort of a bumper on it, like a ribbon about 6–10 inches [15–25 centimetres] wide, and it stood out. It was elliptical the whole way around and the writing was on this bumper. It's nothing like I've ever seen, and I'm an avid reader.[19]

UFO investigator Stan Gordon also located other people who had seen the crashed object. There was a possibility that it was part of the Russian satellite COSMOS 96 which re-entered Earth's atmosphere on that day, but according to the US Space Command records, this satellite re-entered over north-central Canada in the early morning, whereas the Kecksburg object crashed at 4.44 P.M. It seems that the flying object that came down at Kecksburg remains unidentified – but that still does not make it an alien spacecraft.[20]

Even with the accumulated reports of the past forty years, there is still no indication of the origin of the UFOs, and they have not been identified as extra-terrestrial spacecraft. Perhaps we are expecting too much from science, and should turn instead to the witnesses themselves, to see what they can tell us about the objects they saw.

6 UFO entities from our Solar System

Surprisingly often, witnesses have claimed that a UFO has landed and entities climbed out. The appearance and behaviour of these entities vary considerably from one report to another, suggesting that they do not all come from the same place. Nor is it easy to determine the motivation of these alien visitors, or sometimes even whether they are friendly towards the people of Earth. When studied as a whole, the landing reports provide a puzzling hotchpotch of tantalizing yet inconclusive data. Most of the reports do not provide any evidence at all to support the belief that these strange craft come from outer space. But sometimes the entities *do* tell the witnesses that they come from somewhere 'out there', and we will concentrate on these reports in this and the following two chapters. Other UFO entities are less communicative, and for whatever reason they do not give any indication of where they come from. We start this survey of UFO entities with reports of sightings of some of these non-communicative beings, to give a preliminary insight into their activities as observed by Earthlings who naturally placed their own interpretation on what they saw, which may not necessarily be the correct interpretation.

What on Earth are they doing?

A recurring theme is one of avoidance: the entities often do not seem to want to have anything to do with Earth people. If witnesses unexpectedly come across a landed UFO with the entities involved in some activity outside the craft, they quickly get back inside and take off. Sometimes they will even actively prevent the witnesses from getting closer, and this is often achieved by pointing an instrument at the intruder which gives off a beam of light which temporarily paralyses him or her. This happened to José Parra, who was jogging along a road between Valencia and Caracas in Venezuela early in the morning of 19 December 1954. He came across six little men pulling boulders from the roadside and lifting them into a disc-shaped vehicle hovering just above the ground. One of the entities pointed a small instrument at him and he was paralysed by a violent beam, while they hurried back into their craft and rapidly took off.[1]

On other occasions the entities have made attempts to communicate, perhaps using gestures, as happened in August 1958 near Paraibo do Sul in Brazil. The witness approached a craft which he had seen land, and as he got near to it, a porthole opened and a smiling man emerged. The witness was so scared he could not run away. The entity, who had long fair hair and looked like a normal, good-looking man, first picked up and looked at a stone, then stretched out his open hand towards the witness. Then he pointed forwards, followed by another gesture. The witness had no idea what he was trying to say, and in apparent disgust or anger the entity went back to his craft, which then took off.[2]

This entity apparently did not try to communicate through speech at all, but others do, and are often not understood. Three children roller-skating in their home village of Pournoy-le-Chétive in France on 9 October 1954, at a time when there were very many reports of UFOs landing in France, saw a 'round shiny machine' come down. A 4-foot (1.2-metre) 'kind of man' dressed in a 'black sack' like a priest's cassock got out and came over to them. He had big eyes and a hairy head, and the children could not understand what he was saying to them, so they ran away. When they looked back, they saw the UFO climbing quickly into the sky. Its departure was also seen by an adult close by.[3]

As with the Venezuelan case described earlier, witnesses sometimes watch UFO entities engaged in some strange activity, though when they realize they are being observed the entities usually retreat very quickly. Their behaviour sometimes appears incomprehensible, but at other times they seem to be taking part in an expedition to explore Earth and its flora, fauna and geology. The entities seen at Marimbonda in Brazil on 6 December 1978 by hydroelectric plant guard Jesus Antunes Moreira had long black hair and wore metallic blue overalls. They spoke in an unknown language – until they fetched a black box from their craft, after which Moreira could understand what they said, for they were then speaking in his native Portuguese. When he asked what they wanted, they said they were on a research and study mission. They began to pick up stones, and Moreira objected, so they left.[4] The two small entities seen by Pedro Morais on his farm at Linho Bela Vista in Brazil on 11 December 1954 were in a field of tobacco plants. Morais went over to chase them out; but one made a warning gesture and the other ran towards him, before they grabbed a plant and then re-entered their craft and took off at speed. Morais found no footprints, only the hole where the plant had been pulled up.[5]

Entities seen in Tennessee and New Jersey on 6 November 1957 were apparently interested in acquiring a dog. Twelve-year-old Everett Clarke of Dante, Tennessee, saw a landed UFO, long and round like an elongated egg, when he let his dog out at 6.30 A.M. Twenty minutes later he saw the dog and some others near the craft. He went to get a closer look, and saw one of the people near the craft grab at his dog, which growled and retreated. The man picked up another dog, which bit him and escaped. The man then beckoned to Everett, but he kept his distance. The entities went back into their craft without seeming to use any door: 'It looked like they were walking through glass.' Later a 24-foot (7-metre) long imprint was found in the lush grass where the craft had landed. In the evening of the same day, John Trasco of Everittstown, New Jersey, saw a UFO near his barn when he went out to feed his Belgian police dog. There was also a little man 3 feet (1 metre) or less tall, dressed like a leprechaun in green shirt and green hat. He said in broken English, 'We are peaceful people. We don't want no trouble. We just want your dog.' Trasco yelled at them and they retreated without the dog.[6]

The similarity of these two strange tales, too close in time for the second witness to have known about the first event that same morning, strongly suggests that the same entities were involved both times – except for the striking difference in their appearance. Two men and two women, not looking at all peculiar, were seen at Dante; at Everittstown only one entity was seen, a little man in green, with putty-coloured face, prominent nose and chin, and large frog-like eyes. It is sometimes suggested that the stimulus behind UFO events (whatever it may be) causes the witnesses to see them in a form which they find acceptable or comprehensible, i.e. a flying craft with human-like occupants. This would certainly help to explain why such a great variety of craft and beings is described by witnesses.

The need to find water is sometimes another reason given by UFO entities for interaction with humans, though like many other details of UFO encounters, it does seem an unlikely activity. Maria José Cintra, a woman working at the sanatorium in Lins, Brazil, was involved in such an encounter on 27 August 1968. A noise in front of the hospital woke her, and she found a 'foreign-looking woman' at the door, wearing light clothes and a tight headdress. Her speech was not understood, but she held a bottle and mug and clearly wanted water, which Maria fetched. She said that the bottle appeared to be made of glass, and was covered with beautiful engravings. The woman took the water to a bright, pear-shaped craft waiting on the grass, which then took off slowly in a vertical spiral.[7]

Friendly visitors from nearby planets

While some UFO entities have seemed unfriendly towards humans or simply uninterested in them, others have clearly been eager for a closer acquaintance. Conversations have been held, sometimes in the language of the witness, sometimes via telepathy, and witnesses have been invited into craft for trips into space. Real friendships and repeated meetings have been reported by some of the witnesses, who have become known in ufology as contactees, and their stories will be told in Chapters 10 and 11. Such friendly interactions mostly date from the 1950s and 1960s; in the last two decades the emphasis has changed, and the abduction of people by unfriendly aliens into their craft for physical examinations has become a more frequently reported experience. (Some cases of this type will be featured in the next two chapters.) This apparent shift in behaviour probably tells us more about the psychology of the witnesses and the ethos of our times than it does about the aliens supposedly visiting us from outer space. It is also significant that in the 1950s, the UFO entities who revealed their place of origin usually came from somewhere in our Solar System. This was before Man had ventured into space himself, landed on the Moon, and sent probes to other planets which relayed television pictures of conditions there. As Man's knowledge of the Solar System grew, and the likelihood that intelligent life could exist on the other planets declined, so the origins of the UFO entities became steadily more distant and more obscure, as the cases described in the following two chapters will demonstrate.

There are two possible reasons for this. The cynic would claim that all such cases are hoaxes, and the witnesses realized that to make them more believable, the aliens needed to come from further away, from somewhere about which Man has little knowledge. The other possibility is that the entities themselves are prevaricating: they only *say* they come from a Solar System planet, but their true origin is somewhere else entirely. They too have shifted their home planet further out into the Universe as Man's knowledge of the Solar System has grown, in order to make their stories more believable.

One of the earlier encounters with entities who claimed to come from our Solar System took place in 1897, at a time when there were many sightings of mystery airships in the United States. Occasionally witnesses saw and spoke to their pilots, but W. H. Hopkins had one of the pleasantest meetings with them, early in April in the hills east of Springfield, Missouri. He claimed to have seen a shiny metal 'vessel' on four legs standing on the ground. It was about 20 feet (6 metres) long and had three large propellers. Nearby stood 'the most beautiful being I ever beheld . . . She was dressed in nature's garb and her golden hair, wavy and glossy, hung to her waist, unconfined excepting by a band of glistening jewels that bound it back from her forehead.' Both she and a bearded man of 'majestic countenance' who was sitting nearby were fanning themselves, apparently troubled by the heat. Hopkins made friendly gestures, and 'I asked them by signs where they came from, but it was difficult to make them understand. Finally they seemed to do so and smiling they gazed upwards for a moment, as if looking at some particular point and then pointed upwards, pronouncing a word which to my imagination sounded like Mars.' Shortly afterwards the couple climbed back into their craft and were soon out of sight. Hopkins wrote of his encounter in a letter to the *St Louis Post-Dispatch* and despite the incredible tale he told, the paper claimed that Hopkins 'is a prominent church member, and everybody spoken to vouches for his veracity'.[8]

Another early contact, allegedly taking place only a month after Kenneth Arnold's sighting, happened in Parama, Brazil. The date was 23 July 1947, and José C. Higgins was a topographer at that time working in the fields. A large UFO, 150 feet (45 metres) wide, landed and entities in spacesuits emerged. They were 7 feet (2.1 metres) tall, and they encircled Higgins, making gestures as if they wanted him to enter the craft. Higgins spoke to them, with the aid of gestures asking where they wanted to take him to, and one of the entities drew on the ground a round spot encircled by seven circles. He called the central spot Alamo and pointed to the sun in the sky, then at their craft, then to the seventh circle, which he called Orque. This has been interpreted as meaning that they came from Uranus, the seventh planet in our Solar System. Higgins indicated to them that he wished to fetch his wife to take her along with them, and this ruse worked, for the entities did not stop him when he walked away into the forest. From a hiding place he watched them jumping about and throwing large stones in the air, as well as examining their surroundings, before they eventually climbed back into the craft and left.[9]

Mrs Cynthia Appleton, living in Birmingham, England, with her husband and two young daughters, claimed two encounters with alien visitors within six weeks, beginning on 18 November 1957. She was at home, looking after her children, when a figure materialized by the fireplace. He was tall and fair, wearing a tight garment like a silvery plastic mackintosh. Although she saw his lips moving, she heard no sound, and their communication was telepathic. He told her he came 'from another world', but did not say which one. 'Like yours,' he went on, 'it is governed by the Sun. We have to visit your world to obtain something of which we are running short. It is at the bottom of the sea.' A word like 'titium' was in her mind, which her husband later said might be 'titanium'. The spaceman also told Mrs Appleton that on Earth we are using the wrong form of power; we go 'up' (against gravity) while they go laterally. He showed her the image of a spaceship. Before he left, the visitor said he would return in January, and on 8 January two figures materialized, one being the spaceman she had already seen. On this occasion, they both spoke, in English, and told her they came from a country on Venus. They also said that she was seeing only projections of them, and it would be dangerous for her to touch them. After telling her that there would be much bloodshed and suffering in the future, and they would not come to see her again, they bowed to her, then disappeared. In fact further visits were paid to Mrs

Appleton, and during the September 1958 visit she was told that she would have a boy in May 1959. As predicted, a fair-haired boy around 7 lb 3 oz (3.26 kg) was born to her, a couple of days into June, and she said that baby Matthew 'will spiritually belong to a race who live on Venus'.[10]

Also from Venus was the man who rang the doorbell of Mrs Mary King's north Devon cottage on 25 January 1958. Although it was the middle of the night, she let him in and they had a long conversation. She was not completely unprepared, as she had seen a UFO land near her cottage only days before. The entity's appearance was that of a human person: 6 feet 4 inches (1.9 metres) tall, slim, with shining white hair, tanned skin, and wearing a one-piece brown suit. His conversation was uplifting, as he spoke of universal unity: 'We are all brothers. One Creator, One Father, One Family'; and told Mrs King that our world was given all knowledge, but discarded the good and developed the evil knowledge. But eventually the good will triumph, and when that happens, 'will come a period of love and fellowship and Space and Inter-space travel will be easily possible and many, many, more things will be possible, that you have not dreamed of. When this time comes, we shall be waiting to welcome you with out-stretched arms, as true brothers and honourable children of One Creator-Father.'[11] Mrs King was the mother of George King, who founded the Aetherius Society in 1956 and claimed to receive 'Transmissions from Interplanetary Intelligences'. More details of his contacts with Mars and Venus are given in Chapter 12.

The Martian visitors to Gary T. Wilcox's farm near Tioga City, New York State, on 24 April 1964 were interested in rather more down-to-earth matters. While outdoors spreading manure, Wilcox noticed a shiny object on the hill, and drove up to investigate. The aluminium-coloured object was egg-shaped, about 20 feet (6 metres) long, 4 feet (1.2 metres) high and 15 feet (4.5 metres) wide. Two 4-foot (1.2-metre) entities came out from behind it, carrying trays with grassy sods, alfalfa and leaves in them. They spoke to the witness, saying, 'Don't be alarmed, we have spoken to people before.' Wilcox couldn't tell where the voice came from, and although the entities were human in appearance, with arms and legs, he couldn't see their hands or feet, nor could he see any faces. They told him, 'We are from what you know as the planet Mars.' Wilcox and the voice talked for a long time, about manure and fertilizer and growing crops, and they said that their soil was rocky and not suitable for cultivation.

Wilcox asked to go back with them, but they refused to take him, saying the atmosphere was too thin. They only came to Earth every two years, but they have been watching us. They felt that the Earth and Mars, and other planets 'might be changed around'. There was a change taking place in the gravity pull. Then they took off. As they had expressed an interest in using fertilizer, Wilcox fetched a 75-pound (34-kilogramme) bag and left it by the landing site. Next day it had gone.[12] The entities Wilcox met were unusually talkative, and it is puzzling why these sample-gatherers should have wished to communicate, while others seem intent on avoiding humans at all costs. Perhaps the Martians are more friendly than visitors from some other places?

The Plutonians encountered by Mrs Lainchbury seemed friendly enough, however. The events began in the spring of 1964 at Mrs Lainchbury's home in Little Lever, near Manchester. She awoke one night to find her bedroom lit by an orange light coming from a fiery sphere which exploded outside. Two months later, a 5-foot (1.5-metre) entity appeared one night in her bedroom. He said he came from the 'ship' she had seen, and was now stranded on Earth. Further visits followed, and the entities she saw said they were from Pluto. All communication was telepathic, and the letters PLUTO materialized in the air in front of the witness. At the time she did not know this was the name of a planet. Three years later, Mrs Lainchbury saw an orange sphere outside her window, and believed it to be the Plutonians telling her that they were returning home.[13] If these entities really were stranded on Earth for over three years, then they perhaps made contact with others from the band of 'Spacemen on Earth' who are described more fully in Chapter 13.

Written messages from spacemen

John Reeves' encounter with a Martian spaceship and entity has been denounced as a hoax in some quarters, but we cannot confirm this, and the case has some intriguing features. Both Martian and craft were certainly very different in appearance from those seen by Gary Wilcox

*The Martian message
found by John Reeves
after a UFO landing
in Florida in 1965.*

in the previous year. John Reeves lived near Weeki Wachee Springs, Florida, and his first encounter occurred on 2 March 1965, while he was walking in the woods. He saw a landed UFO and hid in the bushes. An entity was walking around, and came within 15 feet (4.5 metres) of Reeves. He saw the hidden man take something from his left side and lift it to chin level; then there was a flash. Reeves guessed he had had his photograph taken. The entity took another before walking back to the craft and climbing inside. Reeves described the 'robot' as follows:

> He was about five feet [1.5 metres] tall. He wore a grey-silver suit, some kind of a canvas suit, and on his head he had a glass dome, sitting right up on top of his shoulders, and I could see right inside that . . . The skin was a very dark skin . . . But his eyes was further apart than a human being. His nose and mouth was the same as ours, but the chin was a little bit pointy. He had some kind of a thing up over his head, something very thin that went over the top of his head to cover his hair. It came down past his ears and you could see his eyebrows. I called him a robot because he is from outer space, but he was a human being just the same as we are.

After the craft had taken off, Reeves found two pieces of folded paper like tissue paper where it had stood. They bore strange writing like 'oriental writing or shorthand writing'. This was later deciphered, and

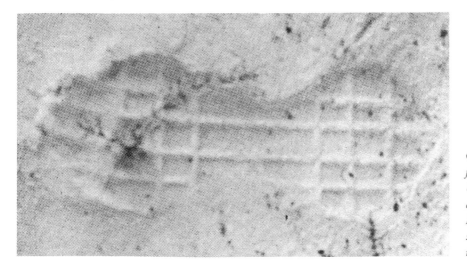

One of the 'alien footprints' seen by UFO investigators alerted by John Reeves after his second UFO sighting in Florida.

John Reeves constructed this replica UFO after his two close sightings near his Florida home in the 1960s.

it read: 'Planet Mars – Are you coming home soon – We miss you very much – Why did you stay away too long'.

On 4 December 1966 Reeves experienced his second UFO landing, this time near his house. His dog alerted him, and outside he saw a craft with its lights blinking. He called UFO investigators, who came out immediately, and they found 'alien footprints' in the sand, together with holes left by the four legs of the craft's landing gear. The prints were followed for at least a quarter of a mile (400 metres), and they were 11 inches (27 centimetres) long and dumbbell-shaped. According to the investigators, similar tracks were left at the site of the first landing the previous year. Strange metal objects were also picked up, and on analysis these were found to contain titanium. Perhaps the Martians were also seeking titanium, like the Venusians who spoke to Cynthia Appleton in the 1950s.[14]

An intriguing footnote to this case is provided by Dr Berthold Eric Schwarz, who reported that in 1957 (eight years before the first Reeves contact) a nine-year-old girl without warning scribbled indecipherable script on the nearest piece of paper. Her father saved the paper, and ten

82

years later came across a magazine with pictures of the Reeves space message. He was shocked to realize that the first thirteen characters of the message were exactly the same as his daughter had written.[15] Dr Schwarz also interviewed John Reeves after his UFO experiences and became aware of the complexities in the case, realizing as a psychiatrist that to label such a case as a hoax was far too simplistic an answer.[16]

There are not many UFO events where as much physical evidence is left as the writing, footprints and metal objects in these two cases. There have been a few other instances where written messages were found, but these are usually dismissed as hoaxes. One such was the case of the Silpho Moor disc, a metallic object 18 inches (45 centimetres) in diameter which allegedly landed on the moor near Scarborough, North Yorkshire, in November 1957. Hieroglyphics were inscribed on the outside, and when the object was cut open a copper tube containing seventeen thin copper foil sheets with more hieroglyphics was found. The outside message read: 'Friends. Message inside to be dealt with by philosophers, not officials. Good wishes, Ulo.' The inside message began:

My name is Ulo and I write this message to you my Friends on the Planet of the Sun you call Earth. Where I live I will not say. You are a fierce race and prepare travel. No one from any other planet ever has landed on earth, and your reports to the contrary are faulty. Men cannot travel far in space vehicles owing to sudden changes in speed direction and many other reasons. They are machines, part at our 'control', part 'auto-control' to avoid objects in way. It is impossible to receive radio over far distances owing to natural waves in space unless key of several frequencies is used, but we can receive single frequencies from near transmitter recorder in space vehicles.

Another correspondent called Tarngee continued the message, which was very long. She included warnings about atomic weapons.

The message I give you is of the atomic energy for destruction. The bombs which the two main powers made when exploded will destroy most animal life on earth and to make it not for long habit by survivors. Any small argument may result in full war then one side after long war will wish to finish it and use atomic bombs . . . Also testing bombs is dangerous. You say it is no danger at present rate after thirty years but rate increases and that time is short. Our recording vehicles on return are sometimes found to have atoms radiating and we can't touch except by machine.

Although the case is generally accepted to be a hoax, it seems an elaborate and purposeless one, and the perpetrator has never been identified. The translator of the message, Philip Longbottom, who devoted many hours to the task, concluded:

The whole thing is not just a simple substitution code, but is a very complicated effort. To make up a complete 'language' like this would seem to be out of all proportion to a hoax, however elaborate. Like any other translator, one tends to get 'inside' the thoughts and feelings of the person who wrote the original, and I firmly believe that this is not a 'made-up' language, but one in constant use. The whole thing flows so easily, and yet contains the natural mistakes that one would expect, considering the difference between our written and spoken word.[17]

Going further back in time to 1954, there is on record from that year another instance of a mysterious written message left after a UFO was seen to land. The events took place early in November that year, at Dudzele in Belgium. A Mr Wardon was returning home on his motorcycle that night when he saw a light coming down. The object itself was 'the shape of a shed' and 10–13 feet (3–4 metres) in diameter, 16–19 feet (5–6 metres) high. It shortly took off again, and at the spot the witness found a small metal box inside which was a parchment with writing in an indecipherable script resembling Hebrew.[18]

Two years earlier there had occurred the event which was the catalyst, if all subsequent messages are indeed hoaxes. Contactee George Adamski (about whom there will be more in Chapter 10) claimed that the Venusian he met on 20 November 1952 in the California desert asked for one of his photographic plates in its holder, and took it away. It was returned to him on 13 December, and strange writing had been substituted for the image originally on the plate. The

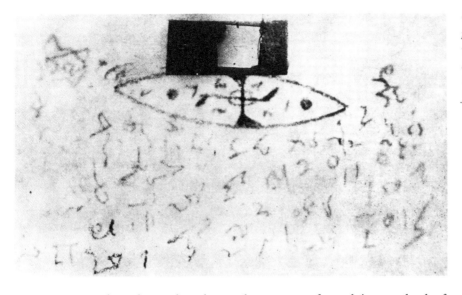

George Adamski's photographic plate with a message said to have been put there by his Venusian friends.

message was thought to be about the spacecraft and its method of propulsion.[19] It was later discovered that the symbols closely resembled primitive characters discovered by archaeological explorer Professor Marcel F. Homet, and published in his book *Sons of the Sun*.[20] Although the book was not published until 1963, the relevant expeditions to the Amazon took place in 1949–50, and some of the designs were published in European newspapers in the early 1950s, from where Adamski could have taken the idea for his space writing, if, as some researchers believe, the space message was a hoax.[21] Adamski and his colleagues also found symbols in the footprints left where the Venusian had stood on the bare desert earth. Plaster casts were made of the best prints, and the two feet are illustrated in *Flying Saucers Have Landed*.

Adamski's widely publicized claims of contact with Venusians had far-reaching effects in the 'flying saucer' world, though many present-day UFO researchers do not believe there was any truth in Adamski's claims. Also believed to be a hoax (though again, as with the Silpho Moor case, not proven and the perpetrators not named) is a UFO contact which has strong links to the Adamski story. The witness was E. A. Bryant who lived at Scoriton on Dartmoor in Devon. On 24 April 1965 while out walking he encountered a UFO hovering close to the ground. One of the three figures that emerged said that his name was 'Yamski', that the entities came from Venus and that they would return in a month bringing 'proof of Mantell'. Later some metal fragments were left at the place, and these were examined to try to ascertain if they really did come from the plane in which Captain Thomas F. Mantell had died in 1948 over Godman Air Force Base, Fort Knox, Kentucky, while chasing a UFO (or a 'Skyhook' weather balloon). The significance of the name 'Yamski' lies in the fact that George Adamski had died the day before, 23 April 1965.[22]

During the same month, an Argentinian shopkeeper named Felipe Martinez had his first encounter with UFO entities, when he saw an egg-shaped craft hovering close to the ground near the Monte Grande. As he ran towards it, he was suddenly paralysed. He watched a small entity about 3 feet (1 metre) tall emerge, wearing a helmet linked to the craft by two cables. The entity, speaking with difficulty, said that they came 'from near the Moon', and were friendly. He also said he would return on 3 May. There was a third encounter on 21 July, and the entity promised a fourth on 3 December. Martinez meanwhile had tried but failed to convince people of the entities' existence, and the entity said they would soon show themselves widely to Earth people. He also promised to take Martinez and his family away to safety before destroying the Earth because people would not accept their existence.[23]

While these entities were threatening to destroy us, those encountered in August 1965 near Mexico City in Mexico were planning a mass landing in October of that year, their intention being to teach us how to use our powers of thought in a more constructive way than at present. Two groups of students, one from the university, the other from a local secondary school, were both independently taken on journeys in the UFO, which was crewed by entities like humans, all being over 6 feet (1.8 metres) tall with fair hair and blue eyes. They claimed to be from Ganymede (one of Jupiter's moons), said that they were a thousand years ahead of us, and knew over 700 Earth languages (so the recording of Earth languages, sent out in the *Voyager* probes – see Chapter 3 – was superfluous!). The witnesses claimed that during their three-hour journey they visited a huge space station, occupied by an assortment of space entities from a number of places in our Solar System. We wonder if the entities 'from near the Moon' were perhaps a break-away faction with some personal axe to grind against Earth people, since their message was rather more belligerent in tone and intention than that of the Ganymede dwellers, who, though intent on conquest, at least claimed to have peaceful intentions. The others claimed to be friendly, yet intended to destroy Earth![24]

A strange case reportedly occurred in 1972 which links back to E. A. Bryant's encounter with the recently dead (or permanently abducted?) George Adamski in 1965. On 22 September 1972, Virgilio Gomez was driving towards the university experimental farm at Palenque, near San Cristobal in the Dominican Republic. A man dressed in green gestured for him to stop, and he did so, thinking he was a military official. Then he saw two other tall beings with yellow-grey skin, dressed in green coveralls, and an oval UFO among the bushes. One entity approached and said his name was Freddy Miller, that he had been rescued from drowning at sea by UFO entities and taken to live on Venus. In fact, a Freddy Miller, who was incidentally a great believer in UFOs, had disappeared on 5 May 1959 while boating on a calm sea. After the entity had talked to Gomez for five minutes (we would like to know what else was said, but our report is deficient in those details), they all left.[25]

During the twenty years since the last case, alien entities have continued to make contact with Earth people, but they have stopped claiming to come from our Solar System. As Mankind reaches out into the Universe, so the entities retreat – and change their tactics.

7 UFO entities from Orion, Gemini, Zircon and other named places

Beginning in the 1960s, intensifying in the 1970s and reaching near epidemic proportions in the 1980s, the trend among UFO investigators for finding abduction cases hidden in apparently straightforward UFO reports has overwhelmed all other types of entity contact case. This trend had its origins in 1961, when Betty and Barney Hill believed themselves to have been taken into a UFO against their will whilst driving home through New Hampshire late at night. They saw a strange light in the sky and stopped to watch as it apparently came closer. Through binoculars, Barney could see figures watching him and in panic he drove off.

Shortly afterwards, Betty began to have nightmares in which she and Barney were taken aboard the alien spacecraft for physical examinations. Barney suffered from insomnia and other symptoms. Eventually they both consulted a psychiatrist, Dr Benjamin Simon, and it was while they were being treated by hypnotic therapy that the full story emerged. Under time-regression hypnosis they relived the events of that fateful night. They described the physical examinations in great detail; and Betty remembered asking one of the entities where they came from. He showed her a 'star map', and asked her if she knew where Earth was on the map. She said she did not, and he replied, 'If you don't know where you are, then there isn't any point of my telling where I am from.' Betty later drew the map while under post-hypnotic suggestion, and in 1968 amateur astronomer Marjorie Fish began the painstaking work, which was to take her five years, of identifying the stars shown on Betty's map. She concluded that it showed stars around Zeta 1 Reticuli and Zeta 2 Reticuli, but another researcher centred the map on Epsilon Eridani and Epsilon Indi. Both interpretations remain controversial, and it is clear that the star map does not provide any proof of an extra-terrestrial origin for the entities who abducted the Hills.[1]

The Hill abduction itself remains controversial. Dr Simon, who conducted the regressive hypnosis, did not believe that the events had actually taken place as described, but once the story received media publicity it set off a chain reaction which is still in motion. Literally hundreds of victims of UFO abductions have come forward, some not even having a conscious memory of the event but being haunted by some strange memory which they wish to explore. When hypnotized these people often produce the by-now-familiar abduction scenario, with a medical examination almost invariably central to the experience. Although some UFO researchers think that the abductions are the key to the whole UFO mystery, others are very doubtful of their value, believing them to be totally imaginary experiences. It is too complex a subject to explore deeply here, and interested readers should obtain some of the numerous books that have been written on UFO abductions, both for and against.[2]

If we assume that the abductions really do occur, then apparently one of the motives is the need to save the alien race from extinction by cross-breeding with humans. This theme has emerged before, most notably in the Antonio Villas Boas case of 1957, which predates all

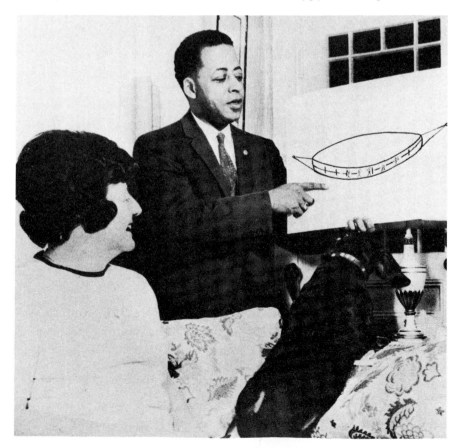

Betty and Barney Hill with a drawing of the UFO into which they believed they had been abducted in 1961.

Antonio Villas Boas (right), who in 1957 claimed he had had a sexual encounter with a woman from another planet.

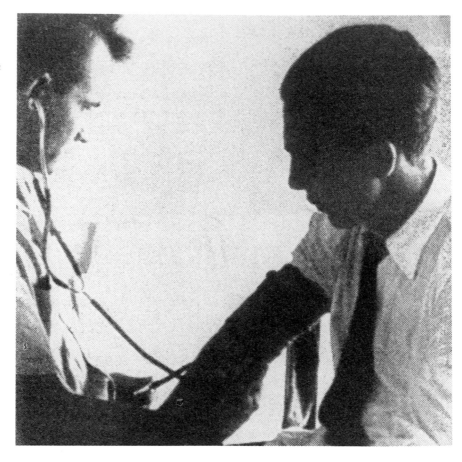

abductions of the 'modern' type. Villas Boas was a twenty-three-year-old farmer living near Francisco de Sales in the state of Minas Gerais, Brazil. He often worked at night, and on 15 October 1957 he was ploughing at 1 A.M. when he saw an egg-shaped machine land in the field. On the day before, and a few days earlier, he had seen strange lights moving around the farm, but this time he found himself close to a very strange spacecraft and his impulse was to flee. But his tractor stalled, and when he tried to run away, he was grabbed by small entities who dragged him into the machine. He was forcibly undressed, sponged all over with a liquid, and had blood taken from his chin through a tube. A while later, a beautiful naked woman entered the room and began to caress him. He found himself responding to her, and blamed this on the liquid which had been rubbed into his skin. Before the woman left, 'she turned to me, pointed at her belly, and then pointed towards me and with a smile (or something like it) she finally pointed towards the sky.'

Villas Boas was frightened that she intended to return and take him away one day, but the investigator suggested she had been telling him that she was going to bear his child on her home planet, and this relieved Villas Boas's fears, since he did not want to be taken to another planet to live.

This case is a fascinating one, rich in detail, for Villas Boas gave a lengthy account of his experience to researcher Dr Olavo T. Fontes.[3] But it is not unique. José Ignacio Alvaro was abducted in a suburb of Pelotas in Brazil on 3 March 1978, and claims that he too had sexual relations with a 'lovely' naked woman.[4] In the following year, on the night of 13–14 April, Jocelino de Mattos was abducted in a suburb of Maringa in southern Brazil, and under hypnosis revealed that he had made love to an alien woman. The entities had also taken some semen from him, and he was told telepathically that they were peaceful and their intention was to study the Earth.[5]

In all these cases the entities did not reveal where they came from, but the assumption was always that they were extra-terrestrial. There was also no explanation of why they wished to mate one of their women with a young, virile Earthman, but it has been assumed that it was to introduce new blood, or to ensure their own survival, or perhaps simply as an experiment. There have also been cases where an Earthwoman has allegedly been impregnated by an alien being, and the resulting baby taken away; details of several abductions apparently with propagation as their motive can be found in Budd Hopkins' book *Intruders*. But we must stress that by no means all researchers accept these cases at their face value, and many instead believe that they have a strong imaginary or psychological component, this belief applying in fact to *all* the abduction cases.

In the 1950s and 1960s the contactees met space people who were wise and benevolent and tried to impart cosmic wisdom to the erring Earthmen; in the 1960s and 1970s the contactees were replaced by abductees, who took part in a scenario possibly more suited to our present technology, with medical examinations and genetic experiments conducted by non-human aliens.

The aliens who abduct Earth people for their own nefarious purposes rarely reveal their origins, though some of them do, and in the rest of this chapter and in the next we will give accounts of a mixture of abduction experiences and 'straightforward' encounters in which the entities have deigned to reveal some information, usually very scanty, about their home base. In this chapter we concentrate on the cases where a name of some kind is given, though not necessarily a name that any astronomer will recognize.

Lonnie Duggan, an Idaho farmer, surprised an alien covered with fur-like hair who was extracting blood from Duggan's horse with a large syringe. The entity spoke to Duggan in a mechanical voice 'like a computer', saying, 'I come from worlds beyond your own. I come from far beyond your Solar System. I come from a planet of the Sun you call Tau Ceti. We are here collecting specimens of the Earth life process.' He also revealed that they first visited Earth over a hundred years ago and have come here quite often since then. They were waiting for the time when Earthmen 'learn to live in peace with one another and to banish wars', and when that happens they will share their knowledge with us.[6]

Josef Wanderka, riding on his motorcycle through woods near Vienna, Austria, in the summer of 1955, encountered a disc-shaped craft in a clearing in the trees. He claims that he rode up a ramp into the machine and talked to its occupants, learning that they came from 'the top point of Cassiopeia'. They spoke fluent German and had a long conversation with him.[7] It is surprising how rarely the people who find themselves in this amazing situation of speaking with beings allegedly visiting from a planet many light-years away, grasp the opportunity to learn something useful. Sometimes they ask where the beings come from, and if it is a recognizable or comprehensible name they remember it; if it is not, they forget it. But they do not often ask why precisely they are here or how the aliens are able to travel such vast distances, which would take more than a lifetime using the methods of space travel currently known to science.

Perhaps we should not be too harsh on them: in the same circumstances would we be calm enough to ask the right questions, and also to remember the undoubtedly complex answers? The contactees' encounters are more satisfactory in this respect: by meeting the entities on several occasions they had more time to think about the situation and to learn something from the visitors. Of course there is no certainty that anything told by space entities to Earth people is true: they may have their own reasons for feeding us 'cover stories'.

An intriguing possibility, for which there is some tantalizing evidence, is that some extra-terrestrials have bases on this planet, in remote mountainous regions such as those in the Soviet Union, Africa and South America, or even under the sea. There have been plenty of reports of UFOs seen diving into or emerging from the sea, and we gathered together enough for a chapter in our book *Modern Mysteries of the World*. Occasionally witnesses who speak to UFO entities are told about terrestrial bases, and one such witness, Olaf Nielsen, claims to have been taken to a subterranean base. While walking near Halmstad in Sweden on 25 April 1960 he was sucked up into a hovering spacecraft and taken to a 'large brightly-lit cavern'. He was shown several UFOs, stored in an underground hangar, and a machine used to create a magnetic protection against intruders. These were not human intruders, but the 'Dark Ones' who came from the vicinity of Orion and would like to take over the Earth.[8]

The entities who took Mario Restier to their planet in 1949 said that they came from near the constellation Orion, but didn't give Restier the impression that they were belligerent towards the Earth. Restier was invited on a trip when he came across the occupants of a landed UFO while driving near Volta Redonda in Brazil on 4 December. For the UFO journey he was immersed in a tub of liquid up to his nose; this was to take away the discomfort of the fast acceleration and to act as nourishment for his body during the journey. The people on the planet he visited were friendly towards him and took him to visit factories. He saw vehicles gliding on suspended roads, and people walking 30 feet (9 metres) up in the air. He also visited a museum where there was a room about planet Earth, known by them as Terra. Thousands of millions of years ago, Earth had been near their planet, his guides told him, but when an enormous celestial body approached, Earth was taken to another solar system. When Restier returned home, he thought he had been away three days, but it was in fact four months by Earth time, and he returned on 14 April 1950. He had another meeting with the Orion people in September 1956, and asked about the time difference. He was told it was to do with the 'Space–Time Contraction' and the 'Synchronism of Time'.[9]

The entities seen by Rubem Hellwig, also in Brazil, may have come from Orion too. Hellwig wasn't sure about the name of their world: they said they came from beyond the farthest star we know, and the name they gave sounded like Arion. Hellwig saw these beings, who acted in a friendly way towards him, in March 1954. At the time he was running a rice plantation near Passa dos Corvos, and encountered the landed UFO while driving towards the plantation. The UFO was small, about the size of a Volkswagen, and shaped like a rugby ball. Two men were visible, one sitting in the cabin, the other outside picking grass. The latter came over to Hellwig, carrying a bottle of reddish liquid, and said in a strange language, which Hellwig surprisingly could understand, that he needed ammonia for his machine. The entity returned to the craft after Hellwig told him where he could get some, and the UFO took off. Next day Hellwig again encountered what appeared to be the same UFO on the same road, but the occupants were different, a man and two women. They said they were scientists, exploring our world, and they showed him their craft, explaining how it worked. They said they would return, and would take him on a trip.[10]

On 7 August 1965 three Venezuelans also met beings who claimed to be from Orion, and again the information they gave the witnesses did not contain any suggestion of belligerence against Earth. The witnesses were two businessmen and a gynaecologist, who were visiting a horse-breeding ranch at San Pedro de los Altos. While out in the pastures, they saw a bright flash of light in the sky, and a huge disc descended, giving off a bright yellowish glow. When just above the ground, and 100 feet (30 metres) away from them, a shaft of light appeared, down which came two tall beings. 7–8 feet (2–2.5 metres) tall, they had long yellow hair, large eyes, and wore shining metallic one-piece suits. They came to within 10 feet (3 metres) of the frightened witnesses and said, 'Do not be afraid; calm yourselves.' Their communication was telepathic, and all three witnesses heard the answers to questions posed by the gynaecologist. He asked the obvious questions first of all: 'Who are you? Where do you come from? What do you want here?'

The entities answered all his questions, and even if the answers were not always comprehensible, it demonstrates that if witnesses can compose themselves and not feel overwhelmed by the situation, they might learn interesting things from the visitors – though we must repeat that it is not wise to place very much reliance on what they say. When all the information from all the extra-terrestrials is gathered together, it fails to provide a cohesive or convincing picture, with few details from report to report corresponding. Yet we should not discard it out of hand: some of the information might be fabricated by hoaxers, some of the information may be intentionally misleading, but some might be the truth. The problem is, how do we distinguish the true from the false?

The story told to the three Venezuelans by the visitors from Orion contained some puzzling information. They said they came from Orion, and were here in peace, to study 'the psyches of the humans, to adapt them to our species'. According to them there are seven inhabited planets, but their list gives eight: Earth, two satellites of Saturn, Epsilon, Kristofix, Kelpis, Orion and a small planet in the Outer Dipper (Ursa Extrema). Their craft are called 'Gravitclides' and operate by means of 'a nucleus of concentrated solar energy which produces an enormous magnetic force'. The beings living on the inhabited planets are not all the same, some are 'Morphous' and some 'Amorphous' (lacking form). The amorphous beings were 'explained' as 'entities, living by means of crisostelic ascending neural evolution'.

When asked if the visitors had bases on Earth, they said that, 'Each planet that sends an expedition to investigate the Earth has a ship almost half the size of your Moon, which they leave behind the planet Mars. This is the reason why more of us are seen when this planet is near to Earth.' They have taken animals, but not humans, from Earth, and they are studying the possibility of creating a new species by interbreeding with humans. (This would explain the activities of the aliens who abduct humans for sexual purposes and medical examinations, and it is interesting that most abduction cases post-date this 1965 encounter.) They did not think much of our 'primitive' space-flight achievements, and claimed to possess a weapon capable of disintegrating the Moon with one discharge. If they used it, life on Earth would come to an end, since our Moon plays a vital role in sustaining life on Earth.

They carried smaller weapons with them to stop plutonium explosions, but claimed to be here in peace. The investigator of this case was an experienced ufologist, and having interviewed the witnesses personally, he felt that their claims should be studied seriously. The witnesses wished to keep their identities secret and were not seeking publicity, in fact their story was kept from the media. One of them was, at the time of the interview, still suffering from shock, and was fearful that the aliens would return.[11]

In 1989 there came scientific confirmation that life is possible in Orion, for astronomer Lucy Ziurys detected molecules containing phosphorus in a large gas cloud in the Orion constellation. William Irvine, the director of the observatory at the University of Massachusetts where the find was made, confirmed the finding meant that 'the basic building blocks for life as we know it are out there'.[12]

The eight inhabited planets named by the visitors from Orion do not include many of the names we have been given by other space beings. What about Mars and Venus? Are they not inhabited after all? Surely the knowledgeable visitors from Orion would know if they were inhabited? Who is lying? The UFO seen by Juan Maldonado in Rio de los Ciervos in Chile, on 16 June 1977, is said to have come from the constellation Gemini. The witness received a telepathic message lasting fifteen minutes, and later by means of ESP the information about the craft coming from Gemini was obtained. Because of its method of reception, this information may be thought less reliable than if it had been received direct from space beings.[13] Another message received without the witness encountering physical entities was that given to 'John' when he was driving along Route 9E in the state of Maine, during the early hours of 14 December 1978. He had felt a compulsion to get up from his bed and drive, and while he was on the road his car went dead. While trying to restart it, he saw a huge dark object hovering above ground. A soft green light shone into the car and John felt as though he were in a trance. The light changed to reddish pink, and it began to pulsate faster and faster. John heard a voice saying, 'Do not be afraid. We will not harm you. We are from the seventeenth star.' After the lights slowed, the voice said, 'We will return'; and then the object left.[14]

'The seventeenth star' is an apparently meaningless phrase; equally illogical (seemingly) is 'a small galaxy near Neptune'. This is where an entity who spoke to Senhora Luli Oswald said he came from. She was abducted from a car travelling along the Brazilian coast near Niteroi, during the night of 15 October 1979. Details of the abduction were revealed under hypnosis, and Senhora Oswald described being medically examined, including a gynaecological examination. One entity told her that some of them originated from Antarctica, and that in Patagonia a tunnel leads under the sea to another world. He himself claimed to be from a different place, the 'small galaxy near Neptune', and not part of the group of aliens that had abducted Senhora Oswald, having been kidnapped himself. When he described what had happened to him, she said it sounded more like a rescue, for he had been travelling alone in a UFO which broke down over the sea, and the others came out of the water and rescued him.[15]

Less than two months later, Elaine Kaiser of Rhode Island was, she believes, abducted from her bedroom. She was levitated up a beam of light which changed colour from white to blue to green. Under hypnosis she revealed that inside the craft she was given a medical examination with two alien beings present. 'I asked, "Where are you from?" He said, "Vector 4 – 2.4 million light-years from here." I asked the little one who he was. He said, "Kelb from Ceres star constellation."' The communication was by mental telepathy, and they also told her that their spaceship was designed for the maintenance and repair of other craft both outside and within the Earth's surface.[16]

This is not the first time we have heard of entities from different places working together, and it suggests a much higher level of co-operation than has yet been achieved among the various races on planet Earth. The names given are, however, improbable. Vector 4 is not a planet we have ever heard of, and the distance of 2.4 *million* light-years sounds equally improbable. Ceres is not a star constellation but an asteroid. The asteroids, thousands of which orbit the Sun between Mars and Jupiter, are very small bodies not large enough to be classed as planets. Ceres has a radius of only 236 miles (380 kilometres).

Equally dubious is the planet Zircon, another unknown name but given by an English businesswoman while under hypnosis following an encounter on 4 March 1982 with an object showing red and blue lights, which hovered close to her car on the lonely A65 road at Coniston Cutting, near Hellifield, North Yorkshire. During the hypnotic regression, the witness did not recall being on board a UFO, but spoke to entities with mythological names like Zeus, who claimed to be from Zircon.[17]

Antonio Nelso Tasca of Chapeco, Brazil, met visitors from 'Agali' during his abduction on 14 December 1983. While driving back to Chapeco at night, he stopped the car and saw a bus-shaped object hovering above the ground. A shaft of light came from the UFO and carried him inside it, whereupon he became unconscious. He was away for ten hours, during which time he encountered 'an enchanting woman' who said she was Cabala from the world of Agali, and they communicated by telepathy. Cabala told Antonio that he had been chosen to receive a message for Earth people, and the message she gave him warned about the dangers of nuclear war. She said space beings are worried about this possibility because it would affect not just the Earth but other worlds nearby, some of them in other dimensions. Extraordinary happenings were due to take place on Earth, 'Masters of the Supreme Wisdom' would come here again and a new society full of light and love would be formed. The message was quite long, but Cabala told Antonio he would be able to remember it. Antonio was reticent about other events on board the craft, but it is suspected that he may have been seduced by Cabala, having a similar experience to that of Antonio Villas Boas and others.[18]

Although the full message does not actually say that Agali is a world in outer space, the message does take a universal viewpoint; however it also speaks of worlds in other dimensions, so perhaps Agali is one such world. It is quite possible, we feel, that many 'extra-terrestrials' do come from worlds in other dimensions close at hand, rather than from places across the Universe many light-years away. One witness, Miguel Herreros, was actually told this. He was fishing in the province of Guadalajara, Spain, in December 1977 when he was invited on board a UFO by strange beings. They asked him questions in Spanish, and told him they came from a place on Earth, but in another dimension.[19]

The message from Agali was concerned with the dangers of nuclear war, as are many of the messages said to come from extra-terrestrials, and it is evident that many of the UFO entities are preoccupied with the foolish and dangerous behaviour of Mankind. As suggested in the Agali message, their concern may not be entirely disinterested: is it not possible that if this physical Earth were destroyed other beings in addition to Mankind would be affected? And yet, although they have a means of penetrating the barrier which separates them from us, they are loath to reveal themselves openly, fearing, probably with good reason, that once able to cross this barrier, the 'developed' nations would set about colonizing, converting, developing and bringing to their world the same blight and devastation that has been spread over much of the Earth. If these theories are anywhere near the truth, the denizens of these other realms face a sorry dilemma: they remain hidden and silent and fall victim to our nuclear insanity, or they reveal themselves and risk becoming victims of our rapacious instinct to conquer. As a compromise, they try to influence us without revealing too much.

Perhaps their understanding of the influence that the average individual has on the policy and actions of governments lacks sophistication. Their chosen method of spreading their warning message seems quite unlikely to succeed if widespread publicity is what they really want: passing the message to ordinary humans, with no influence in the world, is unlikely to result in governments changing their behaviour. A typical example of this involves David Ankenbrandt, who on the night of 29–30 August 1955 was driving home in Cleveland, Ohio. Near Mulberry Corners he saw a bright light descend from the sky, and went to search for the metcorite he thought had fallen. Instead he found 'some kind of aircraft, about 30 feet [9 metres] in diameter, with a dome on top'. When he began to run back to his car, a green light shot out and he found himself paralysed. A tall man, wearing 'something like a ski suit' and speaking English, climbed from the craft and began to tell Ankenbrandt, in a high-pitched voice, not to be afraid, but to tell the government in Washington that 'if there were any more wars here, "they" would have to take over'. Ankenbrandt realized the difficulty he would face, and said so: a 'kid' like himself would not be taken seriously. But the spaceman said he had a week to try to deliver the message, then went back into the craft. Forty-eight hours later, Ankenbrandt returned to the spot and met the UFO entity again, the latter repeating that Ankenbrandt must pass the message to Washington. When he returned with a friend three days later, nothing was seen.[20]

Ten years later, in October 1965, a major power black-out in the north-eastern United States was claimed by UFO entities to have been caused by them, as a warning of what they were capable of doing. Actor Stuart Whitman was in a New York hotel when the black-out happened, and saw two UFOs hovering outside his twelfth-floor window. He heard voices speaking to him in English, as if through a loudspeaker:

They said they were fearful of Earth because Earthlings were messing around with unknown quantities and might disrupt the balance of the universe or their planet. The people in the UFO said the black-out was just a little demonstration of their power and that they could do a lot more with almost no effort. It served as a warning. They said they could stop our whole planet from functioning. They asked me to do what I could to fight malice, prejudice and hate on Earth, and then they took off.[21]

The entity who spoke to a member of the Argentinian Air Force on 20 August 1957 at Quilino in Argentina was specifically concerned with the dangers inherent in the misuse of atomic energy. The witness was somehow prevented from drawing his gun when he saw a UFO landing. He saw no occupants, but a voice spoke to him in Spanish, telling him that they had a base in the nearby Salta region and would soon show themselves to Earth people and warn them about the dangers in their activities.[22]

When UFO entities speak at length to a witness, they often turn to the theme of warfare and pollution. Some students of the phenomenon suggest that the 'message' is not coming from an alien being but from the subconscious of the witness, who instinctively realizes that Mankind's profligate use of terrestrial resources, as well as our too-casual handling of nuclear materials, are leading to a possible future nightmare, a danger that is now generally recognized to be more pressing than ever before. The subconscious chooses the modern image of an all-knowing space being as a way of externalizing this concern. But whatever the source may be, it is a life or death message that must be heeded, for it seems the space people's concern is more than justified.

8 UFO entities from far-distant nameless worlds

Sometimes when Earth people have spoken to UFO entities, the entities have indicated they come from outer space, but the witness has not managed to remember the name given, because it was so strange. Or the entities have not given any name, remaining vague about exactly where they come from. Or they may have used some means other than speech to indicate their origin, such as a diagram drawn on the ground, the meaning of which is usually ambiguous. The entities encountered in December 1952 by a British engineer working in Cape Town, South Africa, were reticent about their place of origin, as if they didn't want him to know. It was dark as he drove in the lonely Groot Drakenstein Mountains, and he was surprised to be stopped by a man waving his arm, who then requested water. The engineer took him to a stream which crossed the road nearby. They used an oil can to carry the water, and the entity took the engineer to a 'strange object' parked off the road. He was invited to enter, and he did so, later describing the interior of the craft and his conversations with its occupants in a long interview with UFO researcher Juan José Benitez. The entities explained how they could conquer gravity, but were not keen to talk about where they came from. The engineer asked, but all the entity said was 'from there', pointing at the stars.[1]

C.A.V., a man who spoke with UFO entities near Lima in Peru, was later hypnotized and asked to remember the name of the entities' planet, which he had been told but had forgotten. All he could come up with was 'Ouja . . . ouja . . . oujan . . . ojan . . . oaja . . . very far away . . . I couldn't understand it.'[2] C.A.V. met the entities among sand hills by the sea, early in 1952. They were very strange in appearance, looking like mummies with joined legs, or one 'double' leg. They wanted to speak to his 'chief', and they were concerned that 'we were playing around with a new substance that could endanger the peace of the world – of the universe . . . they were frightened that, by playing around with atomic explosions, we would create some sort of chain reaction which would not only destroy the world but endanger the universe.' The entities were happy to answer C.A.V.'s questions:

> I asked them who their God was and I noticed a sort of mockery. 'God?' they said, 'What God?' 'Well, the Supreme Creator,' I said, 'who made the universe.' 'Well, we are like gods,' they said. 'How can you be gods?' I asked. And they said that we were very backward. They said that we fought over a flag, over a frontier, that we had wars of trifling matters, like children. We were like children literally still in our diapers. We fought over food, we fought over sex. I then asked them if they did not have those same defects. But they said that they had overcome those years, centuries ago . . . They told me that they extracted from the Sun . . . I learned later that this is called photosynthesis – which I did not know. The sun, which is the chief source of all energy, provided them with all their necessary energy. They obtained humidity from the atmosphere.

When C.A.V. was later asked if their sun was also our sun, he said he didn't know if their planet was in our Solar System or not. He also claimed that they took him for a ride in their craft before departing.[3]

The painter and sculptor Benjamin Solari Parravicini was also taken for a ride in a UFO, being mysteriously transported into the craft he saw above a street in Buenos Aires, Argentina, some time in 1965. As they travelled he could see Japan, France and then Chile, and after one tour of the Earth he was back on the street corner in Buenos Aires. He claimed to have been contacted several times afterwards by the aliens, who told him they were watching over our planet to ensure that there was no catastrophe, but they didn't say where they came from or why it should be so important to them to avert a catastrophe on Earth.[4]

The stories told by the people who have encountered UFOs and their occupants seem to agree in some particulars, but differ considerably in others. For example, the entities often give the same reasons for coming to Earth: either to explore it scientifically, or to warn us against the dangers of pollution and nuclear war; they often say they have been here before and will return; they often invite the witness inside and perhaps give him a trip. But the craft vary in appearance, the entities are almost always different, and the technical information they give about the craft's propulsion is usually different and is generally meaningless in terms of current scientific knowledge. They claim to come from all sorts of different places 'out there', and when witnesses are taken to visit the home planet, their descriptions of what they have seen never agree.

From this we can draw three possible conclusions: there are many different types of aliens coming here, from different places; *or* the witnesses are lying or hallucinating; *or* the entities need to adopt a physical form in order to be seen by humans, and so vary their appearance to appear acceptable or believable to them. They may also be giving the percipient completely false information as well as a totally illusory presentation of their home planet and its civilization. Note how often the percipient views the aliens' planetary home through a viewing screen or window within a spacecraft. This would provide an excellent means of presenting a manufactured 'movie' image rather than any form of reality. Which, if any, of these possibilities is correct remains open to conjecture. It would be pleasing to believe that given a sufficient number of cases the problem could be solved, but we doubt it.

Sidney Padrick of Watsonville, California, was sceptical about some aspects of his own contact experience. He was walking along Manresa Beach on 30 January 1965 when he saw a UFO and was invited aboard. Inside were nine beings in bluish-white suits. Padrick was told where they came from, but 'in terms I did not understand'. During a two-hour conversation he was told about the craft's propulsion system, 'energy transferred through a light source known to them', but he didn't understand what was being explained. When he asked if a scientist could gain more information from them, he was told no, as if they were not keen to give away information about themselves. They said they were here to explore, but Padrick commented, 'I think it was more than that, truthfully. There was a religious facet which I was unable to decipher.'[5]

September 1967 was an active time for the UFO entities in South America, especially for those who wanted humans to visit their world. 'I want you to come with us so that you will know other worlds and you'll realize how small your world is,' said a 'small man' with a large head and prominent eyes who suddenly appeared in Paula Valdez's bedroom on 3 September. She screamed and he fled. This happened in Caracas, Venezuela;[6] in Valencia, Venezuela, the next morning Police Officer Porfirio Antonio Andrade was on duty in City Hall when he heard a humming noise and saw a 4-foot (1.2-metre) tall man with a large head and bulging eyes, wearing a silver coverall. A voice said to him, 'Don't do him any harm. We are here on a peaceful mission. He'll do you no harm.' The voice seemed to come from a disc-shaped craft in the sky, but then the being himself began to speak, asking Andrade in perfect Spanish to go with him to their world, which was very distant and much larger than Earth. But again he was unlucky: Andrade said he couldn't go, being on duty. So the little man flew up towards the UFO, entered through a door which opened, and the craft left. Later in the month, a small entity which could apparently fly was seen in racehorse stables near Caracas, and a trainer was grabbed in bed by strong arms and raked with sharp fingernails (an attempted kidnap?).[7]

Following a UFO sighting early in the morning of 3 December 1967, Police Patrolman Herbert Schirmer 'lost' twenty minutes, and the events which occurred during that time were later retrieved through hypnotic regression. Schirmer had been patrolling on the outskirts of Ashland, Nebraska, when he saw the red lights on an (American) football-shaped craft standing on the ground on tripod legs. Under hypnosis he disclosed that a figure had come out of the craft towards him, holding a 'box-like thing' which flashed a green light or gas over the car. Another entity came forward, touched Schirmer's neck, causing him pain, and then Schirmer got out of the car. The entity asked him, 'Are you the watchman of this town?' When he answered, 'Yes', they took him into the craft. This description doesn't convey the terror Schirmer experienced, which is evident from the transcript of his hypnotic regression session; for example: 'They're getting out. *They're coming toward the car!* It can't be! . . . Trying to draw my revolver. I am being prevented. Something in my mind . . . The one in front of the car is holding up an object . . . stuff shoots out of it and goes all over the car . . .'

They also asked Schirmer, while pointing to the power plant nearby, 'Is this the only source of power you have?' and they asked about the water reservoir. Schirmer described the entities as 4½–5 feet (around

1.5 metres) tall, with tight silvery-grey uniforms, boots and gloves, and a helmet with a small antenna on the side. On the chest was a winged serpent symbol. They had long, thin heads, grey-white skin, flat noses, slit mouths which did not move even when they were speaking, and slanted, unblinking eyes. Inside the craft Schirmer was shown how things worked: the entity pushed buttons and told him things like 'this is an observation craft with a crew of four men'. He spoke through the antenna on his helmet, Schirmer thought, in broken English.

When the hypnotist asked where they were from, Schirmer replied, 'From a nearby galaxy. They have bases on Venus and some of the other planets in our galaxy.' They also claimed to have bases in the USA, one under the ocean off Florida, one in a polar region, and one off the coast of Argentina. All were undersea bases. Schirmer was told that the craft operated 'through reversible electro-magnetism . . . A crystal-like rotor in the centre of the ship is linked to two large columns . . . He said those were the reactors . . . Reversing magnetic and electrical energy allows them to control matter and overcome the forces of gravity . . .' The smaller craft are carried near to Earth in huge 'mother ships' or interplanetary stations, and then released to travel to the bases on Earth. Schirmer was shown pictures on a machine used for surveillance, of 'war ships' flying in outer space, and of the 'mother ship', and he was shown how the entities extract electricity from terrestrial power lines. They also draw power from water. As to their motivations, Schirmer said:

> They have been observing us for a long period of time and they think that if they slowly, slowly put out reports and have their contacts state the truth it will help them . . . They have no pattern for contacting people. It is by pure chance so the government cannot determine any patterns about them. There will be a lot more contacts . . . to a certain extent they want to puzzle people. They know they are being seen too frequently and they are trying to confuse the public's mind. He is telling me they want everyone to believe some in them so we will be open to their invasion . . . He used the word 'invasion' . . . in a friendly way. He said it would be the showing of themselves completely. The public should . . . have no fear of these beings because they are not hostile.[8]

In 1969 in Helsinki, Finland, a man was contacted by entities he could not see, nor did he see their craft. He was at home on the evening of 17 October when he saw a streak of light coiling round his chest. He tried unsuccessfully to wipe it away; then he saw another light in the room. A voice came from above this light, saying: 'We are from another solar system and we have something to tell you . . . We have arrived here on a big craft and at the moment it is behind the planet Mars. For the present we cannot bring it nearer to Earth, for we know that the Earth scientists would observe it with their telescopes, and this would create fear and panic among the people here. At the moment we are in the Earth's atmosphere, having come here in a smaller device.'

The witness tried to persuade the entities to show themselves, but the voice refused, saying it was not possible, but that they would come back in two years, when 'the solar system from which we come, and this one of yours, will be in the most favourable reciprocal position.' The witness should go to a desert and they would join him there.[9] We have no information as to whether this second contact ever occurred, but six months later, another Finnish man met a UFO entity in his home. Kalle Tilhonen lived in Kursu, in a backwoods cabin. On 15 April 1970 he and his small son heard a buzzing noise and saw a UFO outside. Suddenly there was a tall, shimmering figure standing in the kitchen, though he had not come through the closed door. He wore a helmet and had 'big, quite horrible eyes'. The entity spoke of his home 'in the cosmos', and whether Apollo 13 would return safely, before disappearing again.[10]

Are aliens manipulating our world?

During that same month of April 1970, lawyer Raymond Shearer had a UFO encounter while driving home from Madison, Wisconsin. It was late at night, and he first saw a light hanging in the sky. At some point he turned off the main road on to a track, for no reason he can recall, and ahead of him saw a UFO sitting on the road. As he stopped and tried to turn round to get away from it, his car engine cut out. He saw shadowy forms approaching, then blacked out. It was only later, when undergoing hypnosis at a psychiatric clinic after he began to neglect his business and experienced severe changes in behaviour, that the events during two or three missing hours were revealed. Under hypnosis he described seeing the UFO entities gathering specimens – 'rocks, weeds, trees and dirt'. They told him they sometimes also took animals and people, all for their scientists to study. Their home was in 'another galaxy'. One conversation Shearer had with the UFO 'captain' was about time. Shearer had to explain our way of measuring time, and the entity said that time didn't exist: his people could slow time down, speed it up, or stop it. Our ideas about time would prevent us from visiting their home. When travelling through space, they travelled faster than the speed of light. They also used a device that overcame gravity to power their craft. Back on their home planet, they used energy from water, an idea also mentioned by the entity Herbert Schirmer had spoken to a couple of years before. Interestingly, in view of the theory that UFO entities can appear in whatever form would be acceptable to the human witness, the entity Shearer talked to told him that they could turn themselves into a form of pure energy, and go wherever their minds wished to go. Sometimes, when flying in a spaccccraft, they made it invisible so as not to be seen by Earth people.

Several types of space people are visiting Earth, the entity told Shearer, and his people were undertaking a scientific expedition 'to determine the physical characteristics' of Earth. There is also a programme for 'breeding analysis', and some humans have had their brains changed so they are now agents on Earth for the space people.[11]

But perhaps they do not realize that they are acting as agents. Some abductees claim to have had devices implanted into them during the physical examination, and these may be used by the aliens to control the abductees in some way after they have been released. The implications of what is happening are enormous, if we seriously accept all the reports of alien contact and believe all that the witnesses and abductees claim to have experienced and been told. We may be happily living in a world we think we understand, whereas all the time we only see it superficially and are blind to what is really happening. It is possible that, as some esoteric groups have long taught, human history has always been controlled by invisible beings, who come and go as they please and who might as well come from outer space as elsewhere. Is humanity, then, their experiment or a source of energy or simply kept for amusement? We do not know what advantage they obtain for their efforts, but it is disquieting to consider the possibilities – and to realize that there is no one who can positively say that it is not so.

There are certainly many people who claim to have met extra-terrestrials. Remember that the only cases we are including in this book are ones where the entities have indicated in some way that they come from somewhere 'out there'. There are a great many more contact cases where no such claim was made, but where the entities and craft are similar to the ones we are describing. Judging by the number of known reports it would appear that the space people are visiting Earth in great numbers, if, of course, we accept these reports at their face value. The entities who allegedly abducted several members of a rock band who were returning to St Catharines, Ontario, after a party, when asked where they came from, said that it was a long way away, not in our Solar System, and that the questioner wouldn't understand if told. This abduction occurred in the early hours of 16 October 1971.[12]

Time travellers or space travellers?

During the night of 30–31 May 1974, a couple travelling by car from Salisbury (now Harare) to Durban, South Africa, had a strange experience involving a light in the sky, and the journey seemed weird in many ways. They did not recall being taken into a UFO, but when hypnotized several months later, the man, Peter, recalled that they had been 'programmed' inside the car and that the entities in the UFO controlled the car by sending down beams of light. He did not go into the craft, but could see inside it via the light beams. A figure was beamed into the car, and he sat on the back seat throughout the journey. Interestingly, in view of the theory that UFO entities can appear in different forms, Peter said, on being asked what the entities looked like: 'They looked how I wanted them to look.' He said they came from 'outer galaxies', though later he made a contradictory statement, that 'They come from twelve planets of the Milky Way,' which of course is our own galaxy. Another interesting point is that they said they don't believe in God, or gods, a similar statement being made by other space people. They travel 'by time, not by light': 'they are time travellers, not space travellers'. They know all languages of the galaxies, and communicate mentally.[13]

The entities who took John H. Womack into their UFO in Alabama in April 1975 were more than ready to talk to him and tell him about themselves. They spoke at length on life in outer space, and what follows is a summary of some of their information. Their planet is forty light-years away from Earth, but is similar to Earth, only larger. They told Womack:

We know of hundreds of planets in our galaxy which are inhabited by intelligent beings. Because of the great size of our galaxy, known to you as the Milky Way, we have explored only a small part of it. And certainly we are not able to visit other galaxies outside our own Milky Way. But we believe that, at least, thousands of planets are inhabited throughout the Universe. There are many different varieties of intelligent creatures in our galaxy.

They had rescued giants and little bearded people from planets that had been devastated, one by the close passage of a comet, the other by an increase in heat from a dying sun. Of Earth people they said, 'You are the most belligerent and selfish people of all . . . People on other planets, who have mastered space travel, are watching you and are deeply concerned about your ill nature.' But they would not reveal any scientific knowledge, saying that too much knowledge too soon would destroy Mankind. They did reveal, however, that they had not picked up any signals from other galaxies, and did not believe that travel was possible between galaxies: 'We don't believe it possible for a material object to exceed that speed [the speed of light].' However, we recall that entities quoted earlier in this chapter have claimed to travel faster than the speed of light. Womack's entities also believed in a 'Creator' or 'Supreme Intelligence', again contrary to the beliefs of other space beings.[14]

The tale told by Puerto Rican farmer Luis Sandoval is somewhat different from the usual encounter with space entities. On the afternoon of 4 September 1977 the old man was relaxing at home when he heard a noise and saw a flash of lightning 'like a blue candle'. The flash came down near him and turned into 'a little dwarf man about three feet high'. He had long, pointed ears, an ugly round mud-coloured face, big nostrils, a small mouth, and was wearing a jacket and 'little tie'. In a husky voice the being told Sandoval to have courage, and then started to examine him with 'something hanging from his neck, like the thing doctors use when they examine you'. He looked at his feet, knees, chest, back, ears, temples, head and then into his mouth. When he had finished he said he was an extra-terrestrial; then said, 'How nice Puerto Rico is!', before stepping back and turning into a blue flame which shot up into the air and was gone.[15] Perhaps this group of entities was experimenting with physical examination without the bother of abducting people into their craft. Certainly his 'bedside manner' was appreciably more congenial than that usually displayed by the abducting aliens.

In 1978 a Red Army officer, Anatoly Malishev, claimed he had met space people near Pyrogovskoe Lake in the USSR at the end of May or early in June. They took him into their craft and held a long conversation with him, which seemed as if it had lasted three hours, but when he got outside again the Sun and clouds did not appear to have moved. The entities erased most of his memory of what was said, but he did recall them telling him that they do not interfere with life on Earth: they are simply watching us. Malishev was also taken for a ride, first of all to the hidden side of the Moon, where they pointed out their base to him. Then they took him to their home planet, three light-years away from Earth. They landed and Malishev was allowed to get out. He could see no sun in the sky, which was silver-grey and seemed to give off light. The return journey to Earth took only forty minutes.[16]

One puzzling aspect of the UFO contact cases is that usually the entities appear to have no difficulty in breathing the Earth's atmosphere, and whenever humans are taken to other planets they too can breathe the air there without any apparatus. It is unlikely that the air on any other planet is of the same composition as the air we are accustomed to breathe, so does a space journey really take place, or is a false memory implanted by the aliens? There is also the question of alien viruses and bacteria: extra-terrestrials would be very likely to bring new viruses to Earth, and also to become infected by terrestrial viruses from the people they meet. Both humans and aliens would be likely to succumb quickly to new forms of virus, just as native peoples on Earth succumb to diseases to which they have no immunity, when they are exposed to 'civilization'. Yet there never appears to be any precaution taken by the UFO entities to ensure that such diseases are not transmitted. This could well be because they are of a completely different order of life, and our bacteria and viruses can make no impression on them.

The entities who abducted Antonio Carlos Ferreira had procreation in mind, for he was mated with a space woman who was 'chocolate-coloured', like himself. He did not want to mate with her, but the entities took off his clothes and spread oil over his body, after which he had sexual relations with her, the oil probably having acted as an aphrodisiac. This event took place on 18 June 1979, and was one of several contacts claimed by Ferreira, who lived at Mirassol, in Brazil. Interesting features of this case include the fact that some of the alien crew were green in colour, others chocolate-coloured. They wore on their chests the symbol of a cross in a circle. They said they came from another planet, but gave no further details. When Ferreira was abducted some time during 1982, he was shown his child, a daughter named Azelia; she was learning to walk, and physically resembled her mother and the other entities, having pointed ears like theirs.[17]

Abduction of vehicles with occupants

Lorry-driver Harry Joe Turner claimed that on 28 August 1979 he was abducted by a UFO and taken to a planet 2.5 light-years beyond Alpha Centauri, the nearest star to our Sun and itself 4.3 light-years from us. He was driving overnight from Winchester to Fredericksburg, Virginia, when something very strong grabbed his left shoulder. He fired eight shots from his revolver into an 'unseen entity', then blacked out and awoke at the end of his journey, still in his cab but on the passenger side. Although the journey length is 80 miles (128 kilometres), his milometer read only 17 miles (27 kilometres); but the lorry had used enough fuel to do the journey three times.

Turner began gradually to recall what had taken place: both he and the lorry had been lifted into a UFO, where he was examined and questioned by an entity whose speech sounded like 'a tape-recorder played backwards – fast'. He slowed his speech down so that Turner could understand him. Like the others this entity had a series of numbers across his forehead; his name sounded like Alpha La Zoo Lou. Their home planet had dome-covered cities, and Turner thought the place had been devastated by a nuclear holocaust. On the way there, they stopped on our Moon, and Turner saw our astronauts' footprints on the surface. Afterwards he still received messages from the entities, preceded by a ringing in his ears, and he believed that the entities could see through his left eye. He suffered from a 'hysterical personality disorder' following the abduction, and clearly whatever happened to him that night was very traumatic, as is usually the case with victims of alleged UFO abductions.[18]

It is unusual for the witness's vehicle to be taken into the UFO, but Turner claimed his lorry was abducted along with himself, and a woman picked up near Lake Fork Creek in Texas claimed that she and her baby daughter were lifted into a UFO while still in their car. This happened in the early hours of 22 August 1980, while she was driving home overnight. The entities she saw were small with large heads, no ears or eyebrows or eyelashes, no body hair, broad nose and slit mouth. The neck, body and arms were thin and they had only four long fingers. The feet were always hidden by a dense fog. Both the woman and her daughter were physically examined, and she recalled the details under hypnosis. She also remembered being shown seven characters which represented the name of the place they came from: it sounded like Asterisk, but she couldn't pronounce it properly. When she asked specifically where they came from, the entity said they were from the same place as she was, but they had a planet that they go to. Somewhat confusing information, but that may be because we cannot understand what they are trying to tell us. The witness was allowed to ask three questions about the Universe, but the answers again didn't reveal much. The entity did say that they didn't interfere with our lives because we would destroy ourselves: we were a fearful planet with a tendency for annihilation. When the woman asked why she had been chosen, the entity told her that they 'chose samples from the population that already exhibited intelligence and capabilities of understanding properties and theories beyond the present-day time.'[19]

Although the majority of the UFO contact reports in this book describe cases from the USA and South America, they also occur in other parts of the world, and it may be solely the lack of UFO investigators in some countries which has meant that very few cases from certain areas are in the records. Recently an interest in UFOs has developed in China, with the result that some strange cases have come to light. In 1981 a man who was working for a research institute in Beijing and lodging in a hostel there claimed to have had three UFO trips during the first four days of May. During the first trip, a girl in the spacecraft said to him, 'As a reward for your ardour in studying the UFOs we are taking you today for a trip through the immensity of Space.' When they reached what was presumably the aliens' home planet, they were refused permission to land, but below he could see lights and parked UFOs of various shapes. Half an hour later he was back at the hostel.

During the next night he was taken into the craft again and told, 'Today, you will be able to see your own ancestors, and see the original form of the Earth. You will also be able to verify Einstein's theory.' This time they were allowed to land on the alien planet, and he was given a physical examination while there.

On the third night he was taken to the planet again, and during the return trip he asked questions of the blonde space girl. She said they were devoted to peace and did not make open contact with terrestrials because of our wars. Their technology was several thousand years ahead of ours, and they knew our future. The witness refused to reveal what he had been told about the future of Mankind. Six months later, while he was ill in bed, the witness heard a voice telling him, 'We inhabit a planet. We want to exchange scientific information with terrestrials, with a view to raising the technological level of the Earth, for your civilization is too backward in space.' After his experiences this witness suffered mental problems and needed treatment.[20]

Demons in disguise?

Finally, a strange contact case from France which may contain a valuable clue to the real nature of UFO entities. Twenty years ago, we did some research into the similarities between UFO and demon lore, based largely on the book *Demonolatry* by Nicholas Remy, written in the sixteenth century. This contains some intriguing material which inevitably reminded us of the witnesses' descriptions of UFO entities and their behaviour. The description of demons' speech is particularly interesting. When the UFO entities speak to witnesses, the sound somehow doesn't seem to come from their mouth, and they often have surprising command of whatever language the witness happens to speak. Bearing these two facts in mind, read this quotation from *Demonolatry*:

> There are those who believe that certain Spirits, both good and evil, acquaint mankind with a knowledge of the future by means of a voice formed out of the air and sensibly sounding in the ears of men . . . The Demons, without tongue or palate or any functioning of their throat or sides or lungs, inform the air with any speech or idiom they please . . . Witches affirm that their Little Masters speak to them in their own tongue as naturally and idiomatically as one who has never left his native country . . .[21]

Remy could have been speaking about some of the twentieth-century UFO entities.

In the French case referred to, which took place on 12 December 1987 at Malvési near Narbonne, the witness, a composer, was gathering firewood from among the rubble of a demolished factory when he saw six small beings with four machines like snow-scooters. He spoke to them, but they only mumbled in reply. When he said, 'Are you local?', one said two words he could not understand, but later, when mulling over the sounds he had memorized, he thought perhaps they had been 'Planet Earth'. However, this did not occur to him at the time, and he said next, 'So you're extra-terrestrials then?' One of them came forward and drew two signs on the ground, like a lower-case Greek *gamma* and a C, and he said in a nasal tone of voice, '*Ciel, démon*' (sky, demon). The other exchanges, all in French, included:

'Are things better on your planet?' 'Less work.'

'How do these machines work?' 'Magnetic . . . the rain is giving us trouble . . . Not contact with base.'

One held out his hand and said, 'Arctic Pole'; and the woman among them said, 'You don't interest us, but a few humans are departing with us.' Then a seventh entity appeared, sounding angry, and they all mounted the machines which lifted into the air and went off uphill, at a height of 10–15 feet (3–4.5 metres) above the ground. The witness didn't see their final disappearance, as he lost consciousness, though did not fall down.[22]

It is interesting to speculate about what was meant by the entities' cryptic words. They might have been saying that they came from another planet and had a polar base, which would certainly agree with what other entities have said. But if the entity really did say 'Planet Earth' when the witness asked if they were local, then possibly they originated here on our planet – perhaps usually living in another dimension? The use of the word *démon* goes some way towards confirming what some researchers have long suspected: that the UFO entity phenomenon is not peculiar to the twentieth century but has occurred throughout history, the origins and intentions of the entities being understood in accordance with the dominant beliefs of the age.

In other words, they have been with us for a long time: and we still don't really know what their intentions are. Nor do we know if they really come from outer space, as they so often claim nowadays, or whether this is an adaptation to our scientific age with its goal of conquering space. Despite all the information Mankind has been given by the 'space people', passed on by those who have been chosen by them for a contact, we seem to be no nearer to solving the mystery of who they are, where they come from and why they are here.

9 Extra-terrestrial contact by radio

Some of the anomalous radio signals which were picked up in the early years of this century were described in Chapter 3. These were unintelligible signals, and the probability is that some type of natural phenomenon was responsible for them. This explanation cannot, however, account for the strange voices and Morse Code messages which are on record as having been received by various radio enthusiasts over the years.

Messages from Mars?

In 1950 Byron Goodman, a radio ham who was also Assistant Technical Editor of *QST* (an amateur radio publication), wrote of meeting another ham who claimed to have been in radio contact with Mars, Venus and Jupiter since the 1920s. His first contact was with Mars, and the Martian radio ham he spoke to told him how to build better equipment, so that he could also tune in to other planets. He told Goodman:

> Apparently these guys or things on Mars taught the earth language, at least my version of it, to the other planets, and told how to get in touch with me. I figured the whole thing might be a hoax, so I read up on astronomy and darned if everything didn't check. Our dates were made only for times the other planets were visible on this side of the world, and the delay time always checked out on the button. The toughest place to get to was Jupiter, and I finally had to raise my peak power to 200 kilowatts before I could get through, although I'd been hearing them for weeks.

Code was used for these communications, not voice: the beings would not agree to speak with their voices, saying code was 'good enough for all they needed'. Goodman was taken to his shack and they tried for a message there and then. A call came through from 'MM1F' but Goodman was suspicious since there was no time-lag, as there should have been if contact was being made with a distant planet. When he confronted the ham with this, the reply was, 'Oh, that was no planet. That was a mobile station, a space ship practically in our atmosphere. There are quite a few around these days, scouting the earth.'[1]

There are numerous reports of people around the world picking up strange voices speaking indecipherable words. UFO enthusiasts have jumped to the conclusion that, in the absence of any immediately obvious explanation, these must be the voices of extra-terrestrials. Examples include the strange speech heard by radio owners in Spain on 29 January 1950 at a time when UFOs were seen flying over that country;[2] and the chanting voices heard by a Virginia radio operator, beginning on 15 April 1952. These were heard two days running, at 98 mc, and the ham described them as follows:

> The signal was *very* powerful, as indicated by the electric eye, but the voices were low in volume, as if our radio receivers were not precisely designed for the type of modulation being used. The sounds were of even tone like chanting, as if two trumpets were speaking, one low pitch and the other high pitch, like a man and a woman. But the sounds were divided into words and sentences, one then the other speaking. There wasn't a single word in English.

The voices were heard again on 16 March 1953, and this time one word was understood: Washington.

> I made some tests this time, and found the signal to come from a certain direction, *straight to the radio*, not through the aerial. When I stood or placed my hand in a certain place, the signal would dim and static would come in. Evidently this message just happened to affect the FM set on that particular frequency.[3]

A long series of messages

Dr George Hunt Williamson was an anthropologist who had developed a keen interest in UFOs. In 1952 he was living in Prescott, Arizona, and he and his wife and some friends made experiments in communication with extra-terrestrials using, as he describes it, a board marked with the letters of the alphabet, and a glass tumbler to locate the letters and spell out the words of the messages. Using this method, usually called ouija, they received many lengthy communications from entities like Nah-9 of Solar X Group, Kadar Lucu who was 'head of Interplanetary Council-Circle on Mastercraft', Regga of Masar (Mars), Zrs from Uranus, and Touka from Pluto. They referred to their craft as 'Crystal Bells' and talked of landing on Saras (Earth) before too long. They gave the eager sitters information about themselves, their home planets and the Solar System, some of it quite unexpected, like 'Your Sun, which is our Sun also, is not a hot flaming body. It is a cool body.' The heat arose because 'Certain forces come from the Sun and when they enter the earth's magnetic field this resonating field causes friction. And from friction you get heat.' They also spoke about our Moon:

Dr George Hunt Williamson (left), who received regular radio messages from extra-terrestrials in the early 1950s.

All planets are inhabited. Many moons are inhabited also. Planets

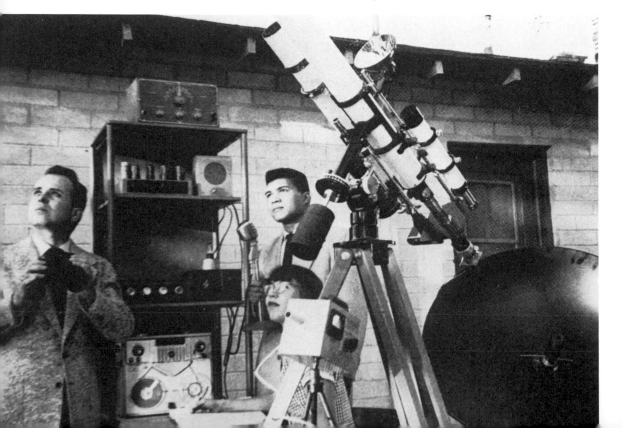

were created to sustain the life of the human race. Your scientists are planning on going into space in rocket ships. You may get to your Moon, but not beyond that. Both of your Moons (yes, you have two; one is the 'dark moon' of Earth. You never see it because of certain conditions) are within your own magnetic field. If you try to leave field with rocket power or atomic power you will be torn to pieces . . . Your first Moon has an atmosphere and water. Some of your scientists have observed snowstorms on the Moon. They have even seen meteors plunge through the Moon's atmosphere. There must be an atmosphere if they see them burn up. There are even inhabitants on the Moon! We have many bases of interplanetary nature there, also.[4]

After a few weeks of receiving such messages by the laborious method of spelling out the words, they made contact with Lyman H. Streeter of Winslow, Arizona, whom they knew as a radio ham. They felt that perhaps easier contact could be made with the extra-terrestrials by radio-telegraphy. Streeter received his first messages via code on 22 and 23 August 1952. As they were not in standard International Morse Code, he found them hard to decipher, and only a couple of words could be made out: Zo and Affa. These were the names of two entities who had already spoken to Williamson; Zo came from Neptune and Affa from Uranus. Streeter later heard that UFOs had been seen from the Lowell Observatory at Flagstaff, Arizona, on 22 August. Over a period of several months, Streeter received numerous messages in code, allegedly from extra-terrestrials. As Williamson commented:

He had been very sceptical at first. However, after he saw discs in the sky where his radio messages told him to look, discs over his own radio antenna, and after messages were received telling about things which no one but he could have known, and finally, messages coming over the receiver that were answers to questions that had never been transmitted to the intelligences in the usual manner, Streeter's attitude changed.[5]

Streeter did in fact secretly test for a hoax: without telling the others, he sent a question on the 40-metre band, and received an answer. Then immediately he switched to 160 metres, and asked another question, to which an answer soon came. No one could have known where he was switching to, and the hoaxer would not have heard the second question, and so could not have answered it. As Williamson noted, perhaps telepathy was being used, so that the entities could pick up the questions without their needing to be broadcast at all![6]

The radio contacts via code continued through 1952 and into 1953. The messages are too long and detailed to be quoted here, but can be read in Dr Williamson's book, *The Saucers Speak*. On just one occasion they heard a spoken message which they believed to be from the space brothers. After a code message telling them to 'vary 92 haste', Streeter

> quickly turned to 92 metres on the receiver. This band is used for aircraft communication. It seemed a speech was being given in a large auditorium. The static was terrible and we could only hear a word now and then. The voice was loud and masterful and spoke perfect English. There was reference to Germany and America and that they could no longer appeal to reason, etc. That is about all we could understand of the message. This was our first and last contact by radio-telephony.

However, from Dr Williamson's description it seems likely that they tuned in to some terrestrial transmission, which coincidentally was being broadcast on 92 metres. If the space brothers were really able to transmit their voices speaking perfect English, surely they would have preferred this method of communication over coded messages? Around the same time, in September 1952, Williamson was making arrangements with the extra-terrestrials to meet them at a prearranged time and place when the craft would land, but unfortunately the two cars they were travelling in were separated and they didn't reach the landing site in time. But Williamson was later on hand when contactee George Adamski talked to an entity from a landed UFO in the Californian desert (see Chapter 10 for further details). The radio communications received by Dr Williamson and his friends ended in February 1953, and after Streeter's death in 1955 Dr Williamson reported that Streeter had believed himself to be a spaceman called Kanet, born on Earth 'to assist the programme of the Space Confederation'.[7]

Another UFO enthusiast who was attempting to make radio contact with extra-terrestrials in the early 1950s was John Otto of California. He tried sending audio messages using light beams as carriers, and he also tried ordinary radio communication. In one experiment in 1954, on WGN in Chicago, which was a 50,000 watt broadcasting station, he asked any extra-terrestrials listening to cut into the WGN transmitter during a certain fifteen-second period. Most people heard nothing, but four listeners reported hearing some sounds, and one recorded something sounding like a short-wave teletype code transmission.[8]

'Ship is real'

In the late 1950s an English ufologist claimed to have received communications, albeit brief, from alien spacecraft on his tape-recorder. Philip Rodgers was a musician and lecturer on music, living in Grindleford, Derbyshire. Despite being almost totally blind, he was a very lively person who did not let his handicap restrict his activities. He is dead now, but will be remembered in British ufological circles for his beautiful guide dog Honey, and for the strange voices he recorded. His technique was to put the microphone on his bedroom windowsill and wait, and in this way a large number of mysterious recordings were made. It all began on 31 August 1957 when he heard 'a long, loud metallic note of 995 cps coming from the sky over Eyam Moor. The object producing it was flying rapidly WNW.'

The recordings were always short, but included 'Hello' in a child's high-pitched voice (there were no children about outside at the time, and anyway no physical voice was heard, it just appeared on the tape), a children's choir, and sounds 'resembling faint Morse signals, strange tremoli, short musical figures, hums, booms, clicks and crashes of a kind I have never heard before.' The strangest recordings were a faint voice saying clearly, 'Ship is real, people', and the 'time-check' recording. Rodgers would always make a time-check when beginning recording, and on one occasion a voice followed this with a twenty-four-hour clock time-check. The voice also made a mistake and then corrected it. Of the strange speech he sometimes heard, Rodgers said, 'In the main they speak a language utterly different from any I know. It has many diphthongs with very few consonants. It is decidedly nasal and is sung rather than spoken.' Their music, he said, 'is different from any I have heard, being thoroughly diatonic but unrhythmic. Phrases I have recorded are distinctly modernistic by our standards.'[9]

No one has been able to explain Philip Rodgers' recordings, but he was convinced that they were the voices of space people. We are reminded of the 'electronic voice phenomenon' (EVP) discovered by a Swede, Friedrich Jürgenson, in the summer of 1959. He recorded birdsong on his tape-recorder, and when he played the tape back he also heard faint human voices speaking in Swedish and Norwegian, and discussing birdsong. He repeated the experiment and obtained voices of entities claiming to be dead relatives and friends, speaking to him personally. In the succeeding thirty years the technique has been refined by other experimenters, but the mystery still remains. Are these really the voices of the dead? The similarity between the EVP voices and Philip Rodgers' voices is compelling, so the question arises: if Rodgers believed he was recording space people, and Jürgenson believed he was recording the dead, perhaps neither was right and the phenomenon has some other explanation. Or perhaps they were both right, and beings existing in other dimensions or on other wavelengths can somehow imprint voices on to magnetic recording tape. How they come through, and what they say, may depend solely on the interests and desires of the experimenter: Rodgers was a musician and ufologist, and he received recordings from musical space people; Jürgenson was interested in birds, and his first recording was of voices discussing birdsong. More recently, messages on a tape-recorder were also obtained by Uri Geller and Andrija Puharich, but as we are describing these radio and tape contacts in chronological order, we will return to the Geller/Puharich voices shortly.

During the late 1950s ham operators were from time to time picking up voices claiming to be extra-terrestrials, like the occasion on 3 August 1958 when radio hams throughout the USA heard a voice on the 75-metre international band. Claiming to be 'Nacoma from the planet Jupiter', he warned that disaster could result from the atomic bomb tests. He spoke for two and a half hours in English, German, Norwegian and his own language, described as a musical gibberish. The signal was powerful, and because of the length of the broadcast, hams called friends to listen in, and Nacoma acquired a sizeable audience.[10]

Edwin, recipient of messages from outer space for the last thirty years.

Meeting the radio contacts

By contrast, the broadcasts from the space brothers received by 'Edwin' had a small audience of only one or two, but continued for at least 30 years. The events began in 1960 when Edwin met a man named George who started work at the same factory, south of Durban in South Africa. One day while the two were out fishing, they began to talk about flying saucers, and George showed Edwin a radio set he carried with him. Soon they saw a light in the sky, which George said was a flying saucer homing in on his radio. Then a voice spoke from the set, saying it was Wy-Ora, commander of the craft they had seen. He came from Koldas, a planet in another universe and part of a confederation of planets. Wy-Ora also said that George was one of them, his real name being Valdar. They wanted Edwin to organize a group to spread information about space beings. Edwin eventually agreed, and was given George's radio when he left to return to Koldas.

Edwin was present when George went inside a UFO which landed on the beach at Richard's Bay. Afterwards, George/Valdar spoke regularly to Edwin on the radio and gave him much information on their planet. For them, time was irrelevant, and they used magnetic streams which crossed the Universe and served as highways. These magnetic streams drove the craft and no fuel was needed. Their universe was an anti-matter universe, a twin of our Universe. Valdar also claimed that their people were trying to civilize the Earth, but there were also other groups of space people on Earth who were trying to undermine their efforts and were malevolent towards Earth.

In 1977 UFO researcher Cynthia Hind met Edwin at his home in Pinetown, and was able to listen to a long message coming live from Commander Herrenoah of Epicot. He claimed to be hovering in a craft 330 miles (530 kilometres) above the house. He said that Epicot was a 'newly found planet' in our Universe, 'a very fertile and beautiful planet'. It was colonized by its discoverers, and the speaker described the problems and what kind of life they led there.[11] In more recent years, Edwin has often received the messages direct: he goes into a trance and the entities speak through him, using his voice. The researcher who has for many years been working closely with Edwin, Carl van Vlierden, has published books on the events: *UFO Contact with Planet Koldas* and *The Twelve Planets Speak*.

Bob Renaud not only spoke to space people on his radio, he saw their faces on his television screen, rode in their spaceships, and visited their underground and undersea bases. The first contact was made in July

During 1961–5 Bob Renaud spoke to space people on his radio, saw them on his television, and rode in their spaceships.

1961 when the eighteen-year-old radio ham was listening in at his Massachusetts home. A high-pitched beeping noise came out of the speaker to be replaced by 'a soft, warm, crystal-clear feminine voice' that asked him to stay on the frequency. She said she was Lin-Erri, and in the months following she gave Renaud much information and advice on what was wrong with our world. The entities came from the planet Korendor, and they told him how to rebuild his TV set so that he could receive outer space TV shows. When he saw Lin-Erri on the screen, she turned out to be a beautiful, shapely blonde, looking about eighteen years old but in reality being seventy-four, 'the prime of life' for the Korendians. They knew Earth as 'Terra', and were involved in Project Terra whose purpose was '1. Scientific and sociological research. 2. Education of the Terrans in the Universal Laws. 3. Prevention of warfare, inevitably to be atomic.' They transmitted a long farewell message to Renaud for the Earth people on 11 November 1965, which ended with these words:

> We are at present engaged in direct encounter with dark forces which control your nations. As each day passes, and they realize more fully that their days on this Earth are numbered, they show their colours more and more. They are in panic now, my brothers.
>
> Your days of sorrow are also numbered, dear friends. Soon, very soon, there will begin a vast change in your lives. You will see the horror of war laid to rest; you will see the ghosts of poverty, hunger and disease exorcized; you will see the fulfilment of your great teachers' predictions. Be assured, that we are working steadfastly with you and for you – to deliver unto you the paradise you have suffered so long to achieve.
>
> Be not despairing, brothers. You are not alone. We are by your side each step of the agonizing way, from the easy path that leads to annihilation to the harder path that curves upward towards the stars. Your destiny is ours, friends. We are with you always.
>
> *Va i amare eno nol si unir.* Farewell for now.[12]

Less exciting in content, but nevertheless dramatic enough for the witness, were the events of 6 September 1967. Dr Edward W. Goldstein was driving home at night to Long Island, New York, when by

Bethpage State Park he saw a cigar-shaped object in the sky. Then he saw a second object, and his car went slower and slower, finally stopping. Then Goldstein heard 'some sort of faint, weird chatter' on his radio – which was switched off. The voices sounded like four different men but he could not tell what was being said. The doctor was forced to sit and listen for half an hour, until the car would start again.[13]

Space people talk to Uri Geller

Uri Geller is best known for his spoon-bending exploits, but in the early 1970s he was also involved in research being conducted by parapsychologist Dr Andrija Puharich, and while they were working together, they appeared to have made contact, via a tape-recorder, with people from Spectra, a distant spacecraft (*not* a planet). They said it was precisely '53,069 light-ages away'. But they would not then explain what a 'light-age' is, though Puharich did ask them. The first contact took place on 5 December 1971 while Puharich was hypnotizing Geller, and then on 9 December they were told that, 'Our computers studied everyone on earth. You were noticed for your abilities as the ideal and perfect man for this mission.' This was to Puharich, and Geller was told that, 'There is a very, very heavy task on your shoulders for the next coming fifty years. There is a lot to be done to help the universe. The cosmic brain will be sent to you.'

Sometimes during these communications they had trouble with the tape-recorder; sometimes tapes vanished; sometimes they were ordered to erase taped transmissions. They were told that the spacecraft Spectra had been 'stationed for the past 800 years over the Earth. It is as big as one of your cities on Earth. But only you can see us.' When Puharich asked if they were also working with other Earth people, the reply was, 'There is no other on earth that we will use for the next fifty years but you and Uri.' On 27 August 1972 the two men were told, 'We hope to land on your planet in a few years. We are seen more and more by people. We will enter your orbital system through [word lost] transformation and be able to enter your environment. You may not understand this . . . One of our failures is that we cannot contact you directly. We can only talk to you through Uri's power on the tape-recorder. It is a shame that for such a brilliant mind we cannot contact you directly.' They said they wanted them to prepare the Earth for a mass landing. They had last landed in South America, 3,000 years before, 'and now we must land again'.

A new voice spoke on 8 November 1972, from Rhombus 4D. He told them that in six billion years our Sun will cool off and Earth people will have to move to another planet. He also spoke at length about Puharich's work with Geller, but we are concentrating here on statements concerning the Earth and outer space. People from Hoova communicated in 1973, saying, 'We first interfered with the human race 20,000 years ago. We came on a planned mission from our own solar galaxy, and our first landing place on Earth was at the place you were at in Israel . . . However, we found traces of the presence of other visitors from other spaces who had been on Earth millions of years earlier.'

They said they had given advice to the ancient Egyptians 6,000 years ago, and also tried to help in Alaska, China and India. Again the message was long and detailed, and on this occasion Puharich was given a definition of a 'light-age': 'one hundred thousand million earth years'. At the end of the communication Puharich asked about reported space visitations, and who these visitors were. The reply was interesting:

> Most of these reports by humans are due to hallucinations and aberrations. But some of our units have landed. But most of the reported landings have been from other visitors from space – some of whom we do not see, but which you can see. They are of different vibrations, different spaces, different velocities. We are the only ones who are mostly here. The other visitors come and go. We stay. This tape will vanish when you hear it. Farewell.

During the two years the contacts continued, Puharich and Geller did not meet any space people face to face.[14]

In the same year that Puharich and Geller's contacts ceased, 1973, an Italian radio ham, Giovanni La Rosa of Gravellona Toca, in Novara province, claimed to have received a strange message while working with his radio. After hearing a peculiar interference, he heard metallic voices speaking Italian. He gave his name, wavelength and position and a voice replied, 'Promise that what you experience from this moment onwards will remain a secret between us and you. Do not speak of it to anyone. We are not inhabitants of your world. We are beings who come from space. We have no intention of doing any harm to your fellow men.' La Rosa asked for a sign so that he could believe what they said, and they told him to be in a certain wood near his home in a week's time. At the rendezvous he saw a circular craft land and three normal-looking people come out, dressed in asbestos suits and helmets with antennae. La Rosa told the beings about his sick child and they promised to come back and cure her, but they never did.[15]

Spaceman interrupts TV programme

In 1977, a 'voice from outer space' broke into a news bulletin on Southern TV, broadcasting across southern England. The drama began at 5.12 P.M. on 26 November 1977, and the voice overrode the television signal for five and a half minutes. The message ran as follows:

This is the voice of Gramaha [alternative renditions include Glon, Gillon, Kilon], the representative of the Asta [or Astron, Ashdown, Ashtar] Galactic Command speaking to you. For many years now you have seen us as lights in the skies. We speak to you now in peace and wisdom as we have done to your brothers and sisters all over this, your planet Earth. We come to warn you of the destiny of your race and your worlds so that you may communicate to your fellow beings the course you must take to avoid the disasters which threaten your worlds, and the beings on our worlds around you. This is in order that you may share in the great awakening, as the planet passes into the New Age of Aquarius. The New Age can be a time of great peace and evolution for your race, but only if your rulers are made aware of the evil forces that can overshadow their judgements.

Be still now and listen, for your chance may not come again. For many years your scientists, governments and generals have not heeded our warnings; they have continued to experiment with the evil forces of what you call nuclear energy. Atomic bombs can destroy the earth, and the beings of your sister worlds, in a moment. The wastes from atomic power systems will poison your planet for many thousands of your years to come. We, who have followed the path of evolution for far longer than you, have long since realized this – that atomic energy is always directed against life. It has no peaceful application. Its use, and research into its use, must be ceased at once, or you all risk destruction. All weapons of evil must be removed. The time of conflict is now past and the race of which you are a part may proceed to the highest planes of evolution if you show yourselves worthy to do this. You have but a short time to learn to live together, in peace and good will. Small groups all over the planet are learning this, and exist to pass on the light of the dawning New Age to you all. You are free to accept or reject their teachings, but only those who learn to live in peace will pass to the higher realms of spiritual evolution.

Hear now the voice of Gramaha, the representative of the Asta Galactic Command, speaking to you. Be aware also that there are many false prophets and guides operating on your world. They will suck your energy from you – the energy you call money and will put it to evil ends giving you worthless dross in return. Your inner divine self will protect you from this. You must learn to be sensitive to the voice within, that can tell you what is truth, and what is confusion, chaos and untruth. Learn to listen to the voice of truth which is within you, and you will lead yourselves on to the path of evolution.

This is our message to you, our dear friends. We have watched you growing for many years as you too have watched our lights in your skies. You know now that we are here, and that there are more beings on and around your Earth than your scientists admit. We are deeply concerned about you and your path towards the light, and will do all we can to help you. Have no fears, seek only to know yourselves and live in harmony with the ways of your planet Earth. We of the Asta Galactic Command thank you for your attention. We are now leaving the planes of your existence. May you be blessed by the supreme love and truth of the Cosmos.[16]

Afterwards there was considerable argument as to whether it would have been possible for amateurs to broadcast this message, though Southern Television themselves were 'reasonably certain that the message was generated by a group of student hoaxers'[17] and we also heard from a BBC engineer that it would have been relatively easy to drive close to the receiving aerials and use a home-made transmitter driven from the car battery to transmit the message. Nevertheless the content of the message is interesting and must have been prepared by someone with a good knowledge of earlier messages allegedly from the space brothers. The serious tone sustained throughout, without any undergraduate humour being slipped in, suggests a serious intent, whatever the original source of the broadcast.

In October 1979 a Peruvian parapsychologist, Caesar Limo, who lived in Chosica, claimed to possess tapes which contained the voices of extra-terrestrials. UFOs had been seen locally, and a press report said of Limo's tapes:

> The primary examination of the tapes indicated that sounds coming from it do not correspond to human voices. It was also verified that the tapes were not run at a speed higher than the normal one; also, it is not tampered with, nor do the tapes reproduce the Morse Code or any other kind of communication signals known on Earth. What is heard resembles faltering monosyllables, as when in certain frequencies radio-telephone conversations are heard over short wave. Experts are convinced they are beings from another world.[18]

Whether all the messages described in this chapter originated with extra-terrestrials, or just some of them, or none at all, is exceedingly difficult to ascertain. The content of the longer messages often makes good sense, and contains the philosophy of the 'space brothers' as revealed in the following chapters, but until incontrovertible proof is forthcoming that Earth people really have been contacted by extra-terrestrials, we must reserve judgement. What is certain is that many people genuinely believe they have experienced these contacts – we do not think they are *all* liars – and perhaps the encounters are indeed 'real', on some level, but not on the physical level we normally experience. The answer may be stranger even than we can imagine.

10 *Contactees meet people from our Solar System*

Some of the most interesting reports of contacts with space people come from the group known as contactees. In previous chapters we have related reports from people who have had brief and usually once-only contact with UFO entities, but what distinguishes the contactees from all of these is the nature of their interaction with the entities. It usually takes the form of an 'ongoing experience', in which the contactee is befriended by the space people, is shown around the interior of their craft, sometimes visits a 'mother ship' orbiting in space, and may even visit their home planet. He (they are usually men) is also given a message of cosmic wisdom, sometimes from a venerable 'master', and may be shown a realistic three-dimensional movie possibly depicting the fate of planet Earth. This can all take place over a period of time and involve many contacts and journeys to outer space. The usual result is that the contactee goes public, with much media attention; sometimes he founds an organization directed towards spreading the space people's philosophy.

A prime example of this scenario was Howard Menger, whose first contact was in 1932 when he was aged ten. On a visit to a favourite woodland glade near his home in New Jersey, he found a woman of unearthly beauty waiting for him. She was clad in a shimmering, glowing one-piece garment and, turning her devastating golden eyes towards him, she smiled and said, 'Howard, I have come a long way to see you, and to talk to you.' She spoke at length about his future and purpose on Earth and told him of further visits he would have from others of 'her people', who, he learned later, came from Venus.

Menger grew up and took part in World War II as a GI in the Pacific campaign. On several occasions during that time he received telepathic messages from his space friends and rendezvoused with one of them at

suitable isolated locations. By 1945 he was out of the army and self-employed as a sign writer, earning a modest living to support his wife and young son. The contacts increased in duration and one task he had was to provide clothing for newly arrived Venusians so that they could move about on Earth undetected. To this end on one occasion he also gave a Venusian male with long blond locks a short-back-and-sides haircut, the better not to attract attention in Main Street.

In 1956, during one of his by now regular contacts, he was able to take some photographs of a saucer landing, but due to its radiations these were technically disappointing. He was also told at this time that the Venusians were going to conduct an experiment to see how people would react 'if people saw this thing over an entire state or even the whole country'. Eventually he was to show his pictures around: 'Tell people we're from other planets and that we are not hostile,' he was told.

On other occasions he met Martians and Saturnians, all of whom could easily appear among Earth people without seeming out of place. In September 1956 he was given a trip to the space people's Moon base, where, with several hundred other Earth people who had been gathered from various nationalities, he was treated to a tour and shown examples of the culture and arts from the various planetary civilizations.

The book Howard Menger wrote about his experiences contains a large section on food, the foods of the space people having a far higher nutritional level than Earth food. A fellow contactee was presented, by the space people, with some of their processed foods and he and Menger had an analysis done of some processed potato, which they found contained about five times more protein than Earth potato.[1] Interestingly this emphasis on food quality is repeated in the experiences of John Watts of Florida, who found that on his visit to Venus the space people were very keen on organic produce and were vegetarians, as were Menger's contacts. Incidentally Watts was the guest of a Venusian woman named Mara, while Marla was the name of Menger's second wife, whom he had known in an earlier life, when he was a Saturnian and she a Venusian. John Watts was also told that there were nine Solar System planets, one permanently obscured by the Sun, and that there are seven solar systems with 200 planets containing life-forms identical to Earth's which are members of the Federation of Planets and are working together to assist Earth's people through the coming cataclysms at the turn of the century.[2]

The Venusians whom Samuel Eaton Thompson met near Markham, Washington, in 1950 were also devoted to spreading benevolence and good will, and like Menger's Venusians lived many years longer than Earth people, due in part to their excellent dietary habits. There, however, the similarity ends, for Thompson's space travellers were simple, ignorant and happy, as they frolicked, unclothed, in the glade by their hovering saucer. Whereas Menger's space people had a high level of technology some 2,500 years in advance of Earth, so he was told, Thompson's Venusians were as children of nature with dark tanned skins, blond hair to their waists; they were also childlike in stature. Naturally they were ignorant of how their spaceship worked or who had made it. It was in fact their home at all times, even on their own planet. Thompson observed that they were without curiosity, any idea of 'sin' or a sense of time – denizens of a golden age one might think, although their opinion of Martians was very poor, rating them as even nastier than Earth people! Like Menger, Thompson had problems when he tried to photograph the spaceship, due to its 'glow' and radiations, and the film turned out blank.

Journeys in Venusian spaceships

Luciano Galli also met Venusians, being contacted in 1957 in the street near his workshop in Rome. A swarthy stranger invited him to go for a ride in a spacecraft, adding, 'Have confidence, nothing will happen to you.' This was not strictly true, for Galli was about to have the most incredible experience of his life, but such was the power of those beady mesmeric eyes that he willingly got into the car and was driven to the outskirts of the city. There they rendezvoused with a flying saucer which took off once Galli was aboard. Through a large, 3-foot (1-metre) wide viewing lens set in the middle of the floor (a standard fitting on many of these medium-sized spacecraft), he watched the Earth recede, appearing to 'look like the Moon' as they shot into space. This sounds disappointingly dull when compared to the beautiful blue-green disc marbled with white cloud formations which is the image of the Earth seen and photographed by NASA astronauts on their Moon journeys and now part of humanity's cosmic imagery.

During the journey Galli chatted with the pilot, who spoke excellent Italian. They then landed on a 'spaceship', a much larger vessel which could accommodate many small flying saucers and much else. When he disembarked inside the hangar he found it crowded with '400 to 500 people' wearing shiny overalls and strolling about talking to each other. They were handsome men and beautiful women, and gave a friendly smile as they passed him. He was given a tour of this vast craft which he was told came 'from the planet you call Venus', then shipped back to his pick-up point, the whole trip having taken three hours and ten minutes. Afterwards Galli denied that the experience was of a mental nature: 'I took this trip in my physical body, this is indeed so. What I say is nothing but the truth.'[3]

Another contactee who met the Venusian 'space brothers' and took trips in their flying saucers came from Sweden and was known as Helge. In December 1965 while he was walking his dog a saucer landed near him and four space people emerged. Wearing nothing but translucent overalls, they presented a strange appearance, being totally hairless with slanting eyes and large pointed ears. They used signs to communicate, and cured him of his painful kidney stones by running a cylinder-like object down his back. Some months later he was contacted again and given the mission of going to the Bahamas to assist the space brothers in their work. He was also given an engraved metal plate and told to wear it always. During the next ten years he made many visits to the Bahamas where other 'brothers' had an underground base. These people were more human in appearance and could easily move among humans undetected, very much as Howard Menger's space people could. He was shown a three-dimensional movie of the Earth's evolution and given various tasks, one of which was to organize a Swedish peace movement called 'The New Generation', but this regretfully did not prosper due to internal dissension. He was also charged with the duties of courier, translator of manuscripts, using codes, and providing information on Swedish military installations. It seems that Helge was never very happy working for them and in spite of their benevolent front was never sure that they were being completely honest with him. He felt he was being used and once said to a friend, 'They are totally without feelings and can witness the most brutal torture. It means nothing to them.' He also remarked that he felt like an animal in their company and thought that perhaps they planned to infiltrate humanity and 'take us over'. On 23 October 1977 Helge died of a heart attack at the relatively early age of sixty-four.[4]

It seems that each group of Venusians behaves differently. Buck Nelson was a farmer living in the state of Missouri when in March 1955 a saucer landed on his farm and the spacemen, naked and carrying their blue one-piece overalls over their arms, strolled into his farmhouse. They shook hands and then got dressed. UFO investigators later told Buck that this was done 'to show friendship and to prove to me that I was talking to real men'. These Venusians, for such they declared themselves, had first contacted him the previous year. Flying low over his house, distracting his animals and blotting out his radio programme with foreign babble, they had used a loud-hailer to talk to him, telling him to raise his arm to answer 'yes' to their questions such as 'Are you friendly?' and 'Can we land safely?' He was also told that there were very

Buck Nelson spoke with naked Venusians who landed on his Missouri farm in 1955.

many Earth people on Venus and that the Moon is uninhabited but contains several underground colonies from various planets. Later he made visits to the planets and returned from one of these with a dog which was nearly hairless. His photographs of the spacecraft were hardly any more convincing as evidence of his experiences, showing only white blobs against the sky.[5]

Contrary to many other contactees' reports, Leo Childers of Detroit, who made his first space flight in April 1955, found that Mars was a dead planet, whereas the Moon contained cities, which were raised and lowered within the craters. Commander Marcosan piloted the craft on this occasion and an 8-foot (2.4-metre) tall crew member 'held me [out of the cabin window?] when I got sick'. This was when they were travelling at 250,000 miles (400,000 kilometres) per hour. That didn't put him off, however, and he claims a total of twenty-one space journeys, possibly a record.[6]

An 'early' contactee, claiming meetings that predated both Arnold's classic UFO sighting and the famous contactee year of 1952 (see later), was Arthur Henry Matthews, who in 1941 was living on his 100-acre

(40-hectare) country property outside Quebec City in Canada. Unable to sleep one night, he took a walk around his buildings and met two 6-foot (1.8-metre) tall men with golden hair and blue eyes who greeted him by name and said, 'We are from Venus and we have come to see what you are doing with Tesla's inventions.' This was a reference to the electrical genius Nikola Tesla who was a friend of Matthews and who had given him plans for unpatented inventions which Matthews was then working on in his workshop. 'How am I supposed to believe you are from Venus?' asked the wary Matthews. The leader replied quietly, 'When you see our ship, you will believe.' He also drew from memory a sketch of Tesla's anti-war machine, which only Tesla and Matthews knew about, and this convinced Matthews that these men were who they said. After showing them his workshop they seemed satisfied with his efforts and then took him to their craft, landed nearby in a low-lying meadow.

It was a huge vessel 700 feet (200 metres) in diameter and 300 feet (90 metres) high, made of what appeared to be a grey metal and shaped like one saucer inverted on top of another. It was known as the Mother Ship X-12. Inside he saw the twenty-four small spacecraft which the X-12 carried and which could be launched from hatches in the hull. Above this the crew had their quarters, which included gardens, recreation area, study rooms and meeting hall. The living quarters were for either single people or married couples, as both sexes were aboard the ship. In the recreation area crew members were playing a ball game and Matthews was once again able to admire their glowing health, with sparkling eyes and blond hair. On a level above this were the gardens where the food was grown and where the crew came to relax and eat, all of their vegetarian diet being eaten uncooked. He also noted that all of the exterior walls were of a translucent material and afforded a clear view of the outside world. Finally he was taken to the control room. Expecting to see banks of buttons, switches and flashing lights, he was amazed to find a bare room with a circular seat in the centre. Four operators sat on this, two men and two women, each one facing one of the cardinal points of the compass. He was told that using only mental powers they could control the movements of the huge craft.

Arthur Matthews had many more visits from the space people during the following twenty years and was taken on a trip to Mars, which reminded him 'of our beautiful eastern townships'. He also met the 'life companion' of the Venusian leader: she had sapphire-blue eyes, golden blonde hair, and a countenance which glowed with an inner spirituality. Standing beside her, the leader told Matthews, 'You may call us Frank and Frances, for we stand for Truth.'[7]

The speed of light is the speed of truth

Another contactee who was impressed, not to say overwhelmed, by the beauty, radiance and perfection of the space brothers and their philosophy, was Orfeo Angelucci. On 24 May 1952 he was driving home in the early hours after completing his night-shift at the Lockheed Aircraft factory in Burbank, north of Los Angeles, California, where he worked as a plastics moulder. For some miles he had watched a hazy red light in the night sky and following it he was led to a deserted area where he pulled off the road and parked his car. The red object shot up into the sky out of sight but as it did so it released two 3-foot (1-metre) diameter fluorescent green globes which floated down in front of him and took up position a few feet in front of the car. A voice, speaking excellent

Contactee Orfeo Angelucci, who in the early 1950s met space people and travelled in their craft.

English, instructed him to get out of the car, which he did, and while he stood by the front bumper the space between the globes became a luminous screen upon which he saw the head and shoulders of a man and woman. Later when describing these beings he used the words 'perfection', 'nobility', 'radiance', such was the impression they made on him. They looked at him and smiled, and although no words were spoken he was aware of telepathic 'thoughts, understandings and new comprehensions' flashing through his consciousness. The figures faded and the voice took over again. Speaking rapidly for several minutes, it gave a brief outline of space philosophy and the concern of the space brothers for the peoples of Earth. It continued with some technical pointers, explaining that these etheric entities had no need of any type of spacecraft, which were only manifest for the benefit of humans, but which could, nevertheless, reach the speed of light. It added, apparently as an afterthought, that the speed of light is the speed of truth.

When Angelucci silently wondered why the space brothers failed to make a massed landing or similar undeniable demonstration of their presence, the voice replied, pointing out that cosmic laws exist which prevent one planet from directly interfering with the evolution of another: 'Earth must work out its own destiny.' Then with the promise of a further meeting, the voice bade him good night.

His second contact came two months later, in July 1952. After having a coffee in a local snack bar he was walking back to his apartment when on one of the vacant lots beneath an elevated freeway he found a remote-controlled spacecraft. Stepping in with confidence, he was whisked up to 1,000 miles (1,600 kilometres) above Earth, in-flight entertainment being provided by the hidden sound system playing, perhaps with a somewhat saturnine humour, his favourite song, 'Fools rush in'. From up there planet Earth had 'a deep, twilight-blue intensity' (not too different from the later photographs taken by Earth's astronauts), though the addition of an 'iridescent rainbow surrounding it made it appear like a dream vision'. He was then subjected to an intense sermon of cosmic philosophy, calculated to make him feel completely unworthy of the great love and compassion that was being shown to him until, he said, 'Tears coursed down my cheeks. Under the spiritual scrutiny of that great, compassionate consciousness I felt like a crawling worm – unclean, filled with error and sin.' They then played the Lord's Prayer as though on a thousand violins and he was zapped by a white beam of light from the ceiling above. At that moment he 'KNEW THE MYSTERY OF LIFE!' and became aware that every-

Son of the Sun, *published in 1959, was one of the books in which contactee Orfeo Angelucci described his meetings with the space brothers and the philosophy he heard from them.*

one was 'TRAPPED IN ETERNITY and ALLOTTED ONLY ONE BRIEF AWARENESS AT A TIME!' Soon he was returned safely to Earth and when undressing later that night he found a circular burn on his skin where the white beam of initiating light had hit him.

In the following months Angelucci pondered how to spread the knowledge of the existence of the space people. Any mention of his experiences to others only produced laughter, sarcasm, ridicule or worse. His wife and two sons suffered nearly as much and it says much for his wife Mabel's strength of character that their marriage survived these traumas. He gave some talks to local groups and some months later self-published his experiences in an eight-page tabloid newspaper format with the title *The Twentieth Century Times*. This brought forth another wave of ridicule and hatred, but Angelucci felt relief in that he 'had satisfied a debt'. However, the reactions to the increasing publicity which his experiences were now receiving were not all negative. Many people read his little newspaper, attended meetings or conventions, and went away quietly questioning whether there might indeed be other life beyond planet Earth.[8]

The most famous contactee of all

A few months after Orfeo Angelucci had his first contact, another Californian claimed to have had similar experiences with the space brothers, and eventually he became the most famous, or infamous, name in the world of flying saucers. He was, of course, George Adamski.

Born in Poland in 1891, he was two years old when his family emigrated to the USA. As a young man he spent some years in the army and later became a teacher of oriental religious ideas using the title 'professor', though he possessed no qualifications nor was he attached to any recognized seat of learning. By 1944 he and some followers were living on the slopes of Mount Palomar, California, where they ran a tourist café. Adamski was a keen amateur astronomer, and in 1946, while observing a meteor shower, he and his companions also noticed a large dark object floating in the sky above the distant mountains, and subsequent news reports of other sightings convinced them that they had seen a spacecraft from another planet. From that moment Adamski

George Adamski (left), American contactee, during an appearance on the Long John Nebel Show.

eagerly watched the skies for further sightings and with the hope that he would be able to photograph the craft through his 6-inch (15-centimetre) telescope. But although he made hundreds of exposures, it was not until 1952 that he succeeded in obtaining the detailed photographs which have subsequently become known worldwide.

His first physical contact occurred in November 1952 in the California desert where with six companions he was on an expedition hoping to see a spacecraft land. Adamski took some photographs of a 'scout ship' hovering about half a mile (1 kilometre) away. After this had left and he was packing his equipment, he noticed a figure in the distance, beckoning him to come over. As he approached he observed the stranger, who appeared to be a handsome man wearing a loose-fitting, one-piece garment of a brown, shimmering material that fitted closely at the neck, wrists and ankles. His long, wavy, sandy-coloured hair touched his shoulders but his tanned face was quite hairless and smooth. Adamski felt that this man never needed to shave. The spaceman calmly observed him with grey-green, slightly slanted eyes and extended a hand in a friendly gesture. Adamski reached out to grasp and shake the proffered hand but the spaceman smiled and shook his head, indicating that they should greet one another by gently placing their palms in contact only.

All the time he was with him, Adamski was aware of 'a feeling of infinite understanding and kindness, with supreme humility' radiating from the visitor, and being 'in the presence of one with great wisdom and much love, I became very humble within myself.' Communication between the two was not easy, and Adamski first asked the stranger where he came from. This question produced a blank response but with further attempts and using sign language and telepathy he was able to establish that the visitor's home was on Venus. He was also informed that the repeated testing of nuclear bombs was causing the planetary people some disquiet and this was the reason why people from many planets, both in the Solar System and beyond, were visiting Earth. The Venusian also conveyed their belief in a Supreme Creator and that their understanding of Him was much broader and deeper than ours, which seems quite an abstruse matter to convey primarily by sign language. After Adamski enquired if they expected a 'life after death', to which the answer was 'yes', the discussion moved on to more factual matters: 'Had any spacecraft crashed on Earth?' – 'Yes.' 'Why don't they land publicly?' – 'This would cause fear and panic among the people.' The Venusian would not permit Adamski to take a photograph, and this Adamski thought was because others of his race who lived on Earth among men could risk being identified.

For very nearly an hour they were together in the desert, until the Venusian indicated that it was time to leave. But before stepping into his scout ship which had been hovering nearby, he requested one of Adamski's photographic plate-holders, indicating that it would be returned before long (see Chapter 6). Then he quickly stepped into the spacecraft which rapidly and silently glided away over the mountain-tops. A mile or so away Adamski's companions had been observing the meeting with the aid of binoculars and had seen the spacecraft leave, so were ready to greet Adamski when he returned. As he walked back, he noticed the footprints left by himself and the Venusian, the latter's being particularly deeply embedded in the sand, and saw that the spaceman's soles had had a pattern on them, which could now be seen imprinted on the ground.

In the following months Adamski met with spacemen in Los Angeles who appeared to be ordinary young men in business suits, and was taken on a visit to a large 'mother ship' orbiting far above the Earth. He was given a tour of the fantastic interior and met and talked with several men and women from Venus, Mars and Saturn. He also renewed acquaintance with his original desert contact, to whom he gave the name Orthon. Other new space friends received the names Firkon and Ramu. Why Adamski should have to give them names might seem rather puzzling, until we are told that space people 'have no concept of names as we use them'. Upon consideration this seems quite reasonable, as we basically use names to identify someone when speaking about them. If instead of having to say Tom, Dick or Harry, we could mentally convey the image and experience of knowing Tom, Dick or Harry, then an audible name would become superfluous for us too. So with his newly named friends, Adamski listened to an hour-long discourse on cosmic philosophy given by an older 'Master of Wisdom'.

On another occasion he was told that on parts of the Moon 'vegetation, trees and animals thrive' and 'people live in comfort'; he was then taken on a trip where from a distance of 40,000 miles (64,000 kilometres) he saw the Moon's surface through the spacecraft's instruments which gave him a close-up view. He saw areas which appeared to be eroded by ancient water flows and a surface that was fine and powdery in some places and similar to sand and gravel in others. As he looked a small, furry, four-legged animal ran across the view, but too quickly to be seen clearly. During another contact he flew around the far side of the Moon and observed, through the viewing screen, snow-topped mountains with timber-clad slopes and mountain lakes and rivers. Here also was a city with people walking the streets and aerial vehicles which travelled above ground. Nearby there was a large hangar, one of many which the space-dwellers built on the Moon to house their visiting spacecraft. After a banquet of exotic vegetables, for these space people ate no meat but had kitchens on board in which they cooked this food, Adamski was treated to a display of scenes beamed directly from Venus. Again he saw mountains, forests, rivers and lakes and various towns and cities of strange but beautiful buildings. He was also shown bathing-costumed Venusians disporting themselves on the white sands of a lakeside. Although all his space friends presented a glowing, youthful appearance, they said that they were many hundreds of years old. Due to their balanced lifestyle, wholesome diet and the benign climate of their planet, the cloud canopy of which shielded them from the Sun's harmful rays, their life expectancy was about a thousand years.

In 1959 Adamski made a world tour, principally financed and organized by his supporters in various countries. He visited New Zealand, Australia, Britain, Holland (where he spent an hour in the company of Queen Juliana and Prince Bernhard), Switzerland and Italy. At every meeting he spoke to an overflowing and enthusiastic audience and had many interviews with the press and TV. Adamski and his group considered the tour of great value in their efforts to spread the message of the space brothers.

Adamski remains one of the major enigmas in the history of the flying saucers. His photographs have never been positively shown to be hoaxes, and those with expertise in such matters admit that they do have the appearance of a large object at a distance from the camera and photographed through a telescope. Added to which, witnesses in other parts of the world have produced other photographs showing vehicles of the same design but from another angle, and the probability of collusion or copying to such accuracy is extremely remote. As with other contactees, Adamski's initial reports and experiences could, just possibly, be factual at some level, in that all these people had an experience which was inexplicable in mundane terms, but nevertheless very real and significant for them. But many commentators observe that in subsequent years conscious invention sometimes took over, as when Adamski reported a journey to Saturn to attend an interplanetary council meeting, perhaps in an attempt to regain the waning public attention to which he had become accustomed.

Also, there always remained the incompatibility of his reports of Venusian and Martian civilizations with the known terrains of these planets, which are quite inimical to any form of life as we know it. The esotericists' answer to this is that we humans can only be aware of the three-dimensional physical level, on Earth and on other planets, and that the entities who contact Earth people normally exist in a dimension of a higher vibration of which we have no cognizance. To themselves, the Venusians and their physical surroundings are completely solid and real, but in order to manifest to Earth-dwellers they must step down their 'atomic vibrations' in order to interact with us. It is interesting to note that some of the entities Orfeo Angelucci met appeared to be quite solid and real, while on other occasions the figure appeared like 'an imperfectly tuned television set' and 'it wavered occasionally, as though I were viewing it through *rippling water*. And the colour did not remain solid and uniform, but varied and changed in spots.' He also wondered if the occupants of the passing cars could see his companion as solidly as he could.[9]

George Adamski, on the other hand, would have none of this. Desmond Leslie, Adamski's British co-author of *Flying Saucers Have Landed*, once asked him, 'Are you sure they did not take on the appearance of our flesh and blood for your benefit?', referring to the Venusians Adamski had met. He received the forthright reply: 'They were not goddamn spooks!' As Adamski appeared obdurate in his views, Leslie did not pursue the matter.[10] But for a man who is said, in his earlier years, to have taught Eastern religious philosophy, such an attitude seems strangely unenlightened.[11]

'We live in peace on Venus'

Orthon, the Venusian whom Adamski met in November 1952, was still on Earth Patrol one morning in November 1968 when he dropped in on Lester J. Rosas, on a beach in Puerto Rico where Lester, a nineteen-year-old student, had been having a swim. Wearing yellow swimming trunks and a blue sweater, Orthon with his long fair hair would have passed for any average hippy of that time. Not that Rosas was completely unprepared for this meeting, as in February of that year Al-Deena, a lovely girl from Saturn, had told him, as they shared an evening snack together in a local restaurant, that he would eventually meet Orthon. Even Al-Deena had not been Rosas' first contact: in 1966 he had photographed a small scout ship as it passed over his home. During this experience he had heard a voice inside his head, speaking Spanish, and had received a telepathic message from Laan-Deeka and Sharanna, the occupants: 'We come from the beautiful planet you call Venus. We have contacted you so that you may help to promote the good work already started by other Earthlings whom we have talked with, both by telepathy and in person. We are human in appearance and our hair is what you call sandy-coloured, like many Venusians'. Our height is about average by your Earth standards. We live in peace on

Venus; there have been no wars on our planet for eons.' In answer to his unspoken question about life on the other planets of the Solar System, they replied, 'Yes, all are inhabited by intelligent beings who are in a more advanced stage of life than yours. We all use telepathy, especially when communicating over great distances, since thought is instantaneous and distance is no barrier to thought communication. But we also speak vocally whenever it is necessary.'

One evening three months later, apparently on impulse, Rosas took a bus ride out of town to a deserted beach area and walked out on to the shore, waiting for he knew not what. Within a few minutes he was approached by a typical Venusian with long fair hair and a one-piece ski-suit. They greeted each other by touching palms. The Venusian identified himself as Laan-Deeka and invited Rosas to walk along the beach a short way. As they did so, the disc-like outline of the Venusian

In 1968 Lester Rosas met the same Venusian who had spoken to contactee George Adamski sixteen years before, both shown in the photograph Rosas is holding.

spaceship materialized. An unseen panel in the vessel's side slid open and out stepped a lovely young woman who said, 'I am Sharanna, Laan-Deeka's fiancée.' She looked about twenty years old, was 5 feet 4 inches (1.6 metres) tall, with blue eyes and blonde hair; Lester's practised Latin eye estimated her vital statistics as 37-27-35. Sensing someone approaching, they entered the spaceship and rapidly gained a height of 50,000 feet (15,000 metres). The Venusians explained some of the technical marvels of their craft and gave Rosas an aerial view of Puerto Rico and San Juan City.

During the conversation Sharanna explained why different contactees' stories varied in their details, saying, 'If their stories are sometimes contradictory, it is with good reason. Your Earth people are contacting space people from different planets and different cultures, and in different stages of advancement . . .' They urged Rosas to publish his experiences and the space message they gave him: that Earth people should cease warring before the Earth is devastated, and stop exploding nuclear weapons which are causing the planet to become unbalanced. Soon they were hovering over the beach again. Rosas said goodbye to his new friends and stood watching the spacecraft rise rapidly into the sky, glowing bright red and making a low humming noise as it did so.[12]

Spaceship from Mercury

When Dan Martin met a pleasant-looking girl from Mercury he was at a loss for words and 'couldn't think of any sensible thing to ask', when she invited his questions. That was in August 1955, when he was driving at night on an isolated road in Texas. Feeling unwell, he pulled over and stopped. As he did so, a 60-foot (18-metre) long craft resembling a diesel locomotive stopped opposite him. He could see some men inside but the lady came out to talk to Martin, because she said she was the only one who could speak his language. With the promise that he would be contacted later and taken aboard a large spacecraft, she stepped back into the strange vessel which took off at a steep angle and disappeared into the night sky.

One evening in June 1956 there was a knock at Martin's door. Outside stood a pleasant-looking man and woman. The man said, 'We have come to take you for a journey on a spaceship. It will be interesting and helpful, I assure you. We will return you to your home within about seven hours, if you are ready to go.' He stepped out between them and as he did so they took hold of an arm each, gripping gently but firmly above his elbows. Immediately the three rose into the air as if in a fast elevator, and he could see the city lights receding rapidly below. Martin became very cold and was wondering how long he could stand it when they stopped below a small craft and entered through an airlock. Ten minutes of flight brought them to a docking with a much larger craft, and once through another airlock he was into the reception area. From here he was taken to meet the captain, who welcomed him aboard the 'Michiel', his ship, which had been on active duty in Earth's atmosphere for over 6,000 years and had been the cause of many of the miracles of biblical and other early records. For one and a half hours the 'captain' told him about these adventures and about Martin's previous incarnations, and then it was time for him to make a tour of the ship.

He was shown rest rooms, bathrooms, a kitchen and store-rooms, with fresh vegetables, fruit, fish and fowl all preserved by a chemical process, not by freezing. After viewing the sick bay and laboratory, he was taken to a dining room where the captain and crew joined him for a meal. Eating started without any preliminary grace or formalities, and during and after the meal he was asked many questions about life on Earth and the customs and practices of humankind, on such matters as the use of money, the making of wars, and religious ideas and practices. As with his first contact, when he was asked if he had any questions he could think of nothing to ask. The party then retired to a lecture room where the captain gave a three-part illustrated lecture, Part 1 on the formation of the Solar System and the introduction of life to Earth, Part 2 on living conditions on the planet Mercury, and Part 3 which Martin could not repeat because he said people would not be able to understand it, as 'it is so radically different from what you are used to'. We suspect that he wasn't able to understand much of it himself either, as by that time he was probably quite exhausted.

He seems to have been singularly uninquisitive when invited to ask questions and the prosaic style of narrative in his published story, coupled with passages of biblical fundamentalism, suggest an unimaginative character unlikely to have made up such an elaborate and detailed adventure story. This gives his narrative an unexpected air of being an authentic experience. After a quick farewell to his hosts he was on his way back to Earth by the same means as he had left it. The final drop from the hovering craft down to his back yard was made alone, and in the grey light of dawn he once again stood on his own doorstep.[13]

Contactee sees spaceship Jesus travelled in

In November 1957 Reinhold O. Schmidt was a salesman and grain-buyer travelling the Middle West area of the USA, when he met with a large silvery craft resting in a dried-up river-bed in Nebraska. As he edged cautiously towards it, full of curiosity, a pencil-width light-beam shone out of the vessel and touched his chest, rendering him immobile. Two men left the ship, searched him for weapons, and then invited him aboard for a few minutes to have a look round. Unlike Dan Martin, Schmidt had plenty of questions, asking the men where they were from, what kind of craft they had and what were they doing, but the leader, who spoke English with a German accent, declined to answer any questions at that time.

Inside the ship were the crew of three men and two women, all unremarkable in appearance having dark hair and tanned skins, and being about 5 feet 8 inches (1.7 metres) tall. They wore shirts and blouses and slacks or skirts. The men were working on the wiring of an instrument panel and spoke among themselves in German, which Schmidt understood, although the leader always spoke to him in accented English. Later he learned that they and their vessel were from Saturn. The vessel itself differed from the disc-shaped 'scout ships' or 'flying saucers' which we have previously encountered being used by the blond Venusians: it was about 100 feet (30 metres) long, 30 feet (9 metres) wide and 14 feet (4.2 metres) high, and of a flattened oval shape with rounded extremities. Like the Venusian ships, although it appeared to be made of a metallic substance from the outside, the walls inside looked glassy and translucent, giving a clear view of the outside world. After being aboard for half an hour, Schmidt was told to leave as the repair work was complete, which he gladly did, being by then secretly worried that he might not get off the strange vessel at all. As he watched, it rose from the ground at an accelerating pace and when about 150 feet (45 metres) high disappeared in a brilliant flash of light.

Schmidt inwardly debated whether he should report the occurrence to some authority or remain silent, finally deciding that 'it was my duty as a citizen to report the whole thing'. He told his story to the Deputy Sheriff who then drove with him to the landing site. They found four imprints in the sand where the landing feet of the craft had stood, and a patch of dark green oil. Later the Chief of Police, the City Attorney and a reporter took him out to the site once more and all agreed something large and heavy had rested there. From then on the situation turned into a circus. The media had a field day with stories of landed spaceships (though Schmidt had never claimed this), hundreds of sightseers flocked into town and the City Police and legal departments took fright and decided they weren't going to authenticate anything. Schmidt was up all night answering phone calls, taking various officials to the landing site and going over his story again and again.

By six o'clock the next morning the police had decided they didn't want a landed UFO on their patch and tried to persuade Schmidt to retract his report. This he would not do, so a lie-detector test was suggested. Not having slept for the past twenty-four hours, he wisely declined, but offered to take one after he had rested if the officials who had originally returned with him to the landing site would take one too. The officials didn't think this a good idea, so instead they placed him under arrest without charge, and he was allowed to go to sleep in the town jail. His next interview was with two Air Force officers who had come from Colorado, and they made a tape-recording of his report. Schmidt was returned to jail and allowed no phone calls or visits from the outside. That evening he was confined in a nearby mental hospital and remained there for two weeks while the doctors did their tests and conducted interviews with him. After his release his employer reassured him that he still had his job and also placed a humorous advertisement in the local press advising corn-growers that 'that crazy grain-buyer is still around . . .' He was deluged with business from farmers who were delighted to be dealing with 'Smitty' again.

During the following three years he was contacted several times by the Saturnians and made some amazing journeys in their craft. His mode of entry became increasingly exotic, ranging from being driven on board as a passenger in their black MG car, to being elevated with his own car as he drove along the highway, up a light beam and then into the spacecraft. As with other contactees, he visited a huge 'mother ship' and there saw a visual display unit depicting the formation of the Earth and its geological history, followed by the history of Mankind. The final scenes showed a civilization on an idyllic Earth – a possibility for the future, if Mankind could avoid its own destruction. Other journeys with his space friends were to the Arctic and below the Arctic Ocean, as their craft was equally at home below the ocean's surface as above. On another occasion the Pyramids of Egypt were the destination, and here the leader took him through a hidden doorway into an unknown underground chamber which housed a 60-foot (18-metre) wide circular spacecraft. Within this was the original cross of the crucifixion, sandals, a robe and crown of thorns, and Schmidt was told that these were the relics of Jesus who, after his earthly death, had travelled back to his own planet, Venus, in this craft.[14]

As with so many other contactees' reports, what started as just possibly credible later degenerated until it became completely unbelievable. If most contactees' experiences do begin with a 'real' UFO experience, which triggers either a totally imaginary 'contact with spacemen' which the witness believes to be equally real, or some sort of genuine contact with alien beings perhaps via another level of consciousness, then once the attention paid to the witness by friends, relatives and the media has died down, the contactee may resort to conscious (or perhaps even unconscious) fabrication of ever wilder stories in order to rekindle people's interest and once again become the centre of attention. Sometimes the friendships with space people also fill voids in the lives of the contactees, as the next chapter will demonstrate.

11   *Contactees meet people from Meton, Bâavi, Clarion, Zomdic and other far-away worlds*

All the contactees whose experiences we related in the previous chapter were men. In fact women contactees are very rare, though they do figure more often as abductees, a category of UFO encounter in which the victims rarely receive any comprehensible information from the aliens as to their origin or intentions. But perhaps the most glamorous, and scarcely believable, contactee was a woman, Elizabeth Klarer, who not only met a friendly and handsome spaceman, projecting warmth, wisdom and brotherly love, but also became his lover and the mother of his child.

Elizabeth Klarer saw her first spaceship in 1917 at the age of seven as it swooped over her home in Natal, South Africa. She became a meteorologist and obtained a pilot's licence, and in 1937, during a flight to Johannesburg with her husband, had another sighting in an air-to-air close encounter. But her story really began later when at her South African farmhouse within sight of the Drakensberg Mountains a flying saucer came in to land and the romance of her life, Akon, stepped out to meet her. The encounter was one of intense emotion; in her own words:

A UFO photographed in the 1950s by Elizabeth Klarer, who claimed a close relationship with space traveller Akon.

'My heart beat against my ribs with suffocating intensity; I felt faint. A man from another planet, another world, influencing my life! Time seemed to stand still at that moment. There was no fear, only a deep and exciting happiness.' (Klarer does in fact seem to live at a fairly high emotional pitch most of the time, as when she tells us: 'The shrill and demanding clamour startled me out of my reverie, the peace and seclusion of my office transformed to bedlam by the screeching monster' – and that's only the telephone ringing.) Overcoming her fit of the vapours she bravely ran up to her tall spaceman standing beside his ship.

> Laughing gaily he caught me round the waist and swung me up on to the hull of his ship and we both laughed as though it was the most natural thing in the world. Then he spoke to me in precise English and his voice was like a caress. 'Not afraid this time?' Still holding me close in his arms, he smiled gently as I looked up into his kind grey eyes. 'I have known your face within my heart all my life,' I answered. 'I am not from any place on this planet called Earth,' he whispered with his lips in my hair.

Which is really all that concerns us right now. Akon was in fact from Meton, a planet in the Alpha Centauri system, the nearest bright star to the Solar System at 4.2 light-years' distance, and he was a research scientist who travelled to many planets for his work. As the spaceship accelerated up through the atmosphere, with Klarer on board, she noted the Earth 'floating in space, her delicate blueness lightly shrouded in white clouds swirling in wind patterns' – this being perhaps the most accurate description of Earth's appearance from space given by any contactee. With the frankness born of supreme confidence, Akon stated, 'We rarely mate with Earth women; when we do, we keep the offspring to strengthen our race and infuse new blood.' Klarer stayed on Meton until her space-son was born, which was achieved without pain or trauma but 'with no effort at all, just a deep feeling of achievement and pleasure' and with Akon acting as midwife. The space baby was named Ayling, and Klarer's book about her adventures closes nineteen years later when the youth and his father occasionally visit her in South Africa.[1]

An even earlier contact with the Alpha Centauri system was made by a Frenchman identified only as Monsieur Y. Some time during the war, in the early 1940s, he had been contacted by space people who had taken him in a spaceship to their planet Bâavi in the Alpha Centauri system. The journey lasted only one and a half hours, the distance of more than 4 light-years being covered so quickly because the spaceship, which he called a vaïd, attained more than 'gravific speed'. This caused them to enter anti-time, changing course three times on the way to Alpha Centauri. On the planet Bâavi Monsieur Y found that men and women had complete equality and lived in a large metropolis, the rest of the planet being left to the natural wildlife. In their early years the young people had children and were then sterilized, living for several hundred years without apparent ageing. They had no marriage or permanent partners but loved everyone equally and practised a form of free physical love among themselves. They foresaw that a great cataclysm would hit the Earth in a relatively short time and so they planned to save some humans to repopulate the planet afterwards.

Perhaps this is what Akon and his friends from Meton were engaged in when he met Elizabeth Klarer. Monsieur Y also returned with much documentation which included a grammar of the Bâavi language, descriptions of intergalactic spacecraft and their construction, a philosophy of time, and the fundamentals of Bâavian science written in Armenian: these last were discovered by Monsieur Y in Southern Algeria buried in a cave and revealed to him by an old desert nomad. Why an archaic Armenian text should be buried in Algeria and contain information on the planet Bâavi, not even Monsieur Y knew.[2]

'You may speak to our captain in our scow'

It is regrettable that Monsieur Y, with so many fantastic experiences and so much amazing documentary material, did not, so far as we know, ever organize it into published form. Those of the contactees who did publish their experiences have produced stories both incredible and fascinating, and Truman Bethurum's report of his meetings with space people in the Nevada desert contains both of these qualities. Bethurum was a practical, down-to-earth and hard-working man engaged in highway construction, and at the time of his first contact was on a

Contactee Truman Bethurum, who met people from the planet Clarion.

project in the Mormon Mesa, Nevada. He was on night-shift mainten-ance work and on the night of 27 July 1952 during a lull in activity had driven into the desert to take a nap, away from the pressures and demands of the camp. Later he awoke to find ten small men standing around his truck watching him.

When he got out he saw, in his words, a 'flying saucer' 300 feet (90 metres) in diameter and 18 feet (5.4 metres) deep. It appeared to be made of stainless steel and was hovering noiselessly several feet above

156

the desert floor. The crew of the saucer had dark olive skins and dark hair and were between 4 and 5 feet (1.2 and 1.5 metres) tall. Crowding round him, they were anxious to shake hands, in a conventional Earth-like manner. 'Where do you call home?' asked Bethurum, and he was answered with a sentence that might have been spoken by one of the little people of fairy tradition: 'Our homes are our castles in a far away land'; the speaker then added, 'You may speak to our captain in our scow', which it seems was their name for their spacecraft. The captain was a little shorter than any of the crew, but more than made up for this by being a delightfully attractive woman with a glowing skin of olive and roses and with brown eyes that smiled and flashed, shrewdly assessing Bethurum for what he was. On her short black hair which was curled up at the ends she jauntily wore a neat black and red beret, and her black velvet bodice and red pleated skirt continued the colour combination. She smiled and invited Bethurum to speak; he stammered, 'Are you from some European or Asiatic country?' She replied, 'We travel interplanetary, and it has been only recently that we have landed on your soil.' When he asked how far they had come and how long it took, she said, 'Time and distance are of no concern to us, and what you call time and distance is inconsequential in our lives.'

It wasn't until the third visit of the scow that he learned the name of the lady captain, which she gave as Aura Rhanes. She was from the planet Clarion which was in the Solar System but was never seen from Earth because 'it was on the other side of the Moon'. This apparent nonsense was later rationalized by Bethurum, saying that he believed Clarion was in the same orbit as Earth but always behind the Sun, and so never seen from Earth. 'Mars is a beautiful place to see,' she told him. 'Yes, there are people there, just like you and me,' and all the planets had an atmosphere similar to Earth, in fact many planets were inhabited.

When Bethurum spoke of his experiences to his friends and work-mates he met with the usual ridicule and hostility, and also threats to his person from some of the more belligerent gun-toters who believed he could be meeting 'Commie' agents out in the desert. His wife, who lived at Redondo Beach, California, with their children, was quite antagonistic to his stories, and feared that his work in the desert heat had somehow unhinged his mind.

During the following three months Bethurum met the lady from space eleven times, each time out in the desert at night but within her landed spaceship. He was never allowed to use a camera and was never given any artefact as evidence, though he once gave her a ballpoint pen she had borrowed. Each meeting was a leisurely one and they spent long hours together talking of conditions on their respective planets. Naturally Clarion was unlike Earth, for there peace, prosperity, well-being and love prevailed, and though they enjoyed a high level of technology it was directed towards maintaining a natural and simple style of life. Aura Rhanes several times hinted that Bethurum and some friends might be able to take a trip to Clarion one day, and at the tenth meeting he eagerly listed those friends who would come with him. Aura Rhanes explained how they would spend their time on Clarion. But of course it never happened. After one more visit Aura and her spaceship failed to return. Bethurum spent many a solitary night scanning the desert skies and sometimes lighting a flare as a signal to the spaceship, hoping that the lovely lady captain would return, but she never did.[3]

Contact with Coma Berenices, Zomdic and Shebic

Paul Villa was another mechanic who met space people; this was on 16 June 1963 near Albuquerque, New Mexico. He had received a telepathic message to rendezvous with the landed craft and met the group of five women and four men when they came out to see him. They were from 7 to 9 feet (2.1 to 2.7 metres) tall and their hair varied in colour: black, blonde and red. They were, they said, from the galaxy of Coma Berenices. This was no more informative than most other such pronouncements. Coma Berenices is a star group north of Virgo and within that area of sky a number of galaxies are visible, but there is no galaxy known as Coma Berenices. When they took off again they obligingly hovered overhead while Villa took some clear, well-exposed colour photographs. Unfortunately they were so good that they were instantly dismissed as fakes: such are the trials of a contactee. Villa is now one of that group of evangelizing contactees who prophesy Armageddon. Having produced many UFO pictures, he distributes them to everyone from heads of state to the very humblest, along with the message that UFOs are only 'a small part of God's huge armies that will soon invade planet Earth and redeem humanity from their present immoral fallen condition.'

Another engineer, this time in England, was also contacted by telepathy before the space people landed and took him aboard. He was James Cooke from Runcorn in Cheshire and at two o'clock in the morning of 7 September 1957 he watched as a 120-foot (36-metre) diameter spacecraft glowing with a spectrum of colour homed in on him. A ladder came down from the hovering vessel and a voice told him, 'The ground is damp. Jump. Don't have one foot on the ground or you will be hurt.' So he jumped and successfully entered the craft, which appeared to be unmanned.

Soon he was directed to leave the vessel and found he was within a larger spacecraft where he met beings who were taller than the average Earth person, and used telepathy for communication. He described them as beautiful, with smooth 'baby-faced' skins, though some did have beards. Later he realized that they were neither men nor women, but that everyone was bisexual. He observed their surface transport system, which was operated by musical vibrations, and learned that as they could directly turn energy into whatever form they needed, they had no use for money.[4] He was in fact on planet Zomdic 'in another solar system' which was unknown to Earth astronomers. He met a planetary master who told him, 'The inhabitants of your planet will upset the balance if they persist in using force instead of harmony. Warn them of the danger.' 'But they won't listen to me,' Cooke pointed out. 'Or anybody else either,' sighed the ancient one. Cooke arrived back home at 10.50 on Sunday evening, having spent forty-four hours on his adventure.

Two years later he was off again, this time to planet Shebic where the days were longer and the nights colder than we experience on Earth. Here the people had an average height of 5 feet 2 inches (1.5 metres) and golden brown skins. They had no industry, and how they obtained their technical necessities was not explained. Afterwards Cooke appears to have made no attempt to go public with interviews, lectures or books, as other contactees have done. Does this make his claims more believable? Or was he failing the space people in their plan to awaken Mankind?[5]

Telephone messages from Aenstrians

In the quiet country town of Warminster, Wiltshire, a local journalist, Arthur Shuttlewood, held court as flying saucerer for many a year. With his books and articles chronicling the innumerable lights and noises which beset the town from the early 1960s, he provided a focal point for the enthusiasts who sought a Mecca at which to meet other devotees. For years Shuttlewood had been only a puzzled and honest investigator of inexplicable local reports, but in September and October 1966 he received a series of strange phone calls which firmly placed him in the contactee category.

The callers identified themselves as Caellsan, Selorik and Traellison, all from the planet Aenstria. Their message for Mankind and the information regarding themselves and their home planet was very much the same as given to other contactees. They said that the repeated testing of nuclear weapons was polluting air, earth and sea and would

People from the planet Aenstria made contact with English ufologist Arthur Shuttlewood in the 1960s.

soon begin to poison life on Earth, and they were removing some dumps of radioactive materials from the sea bed for safe disposal. They were here to save Mankind from destruction and to advise us to seek enlightenment and truth: 'Learn to live fully and not fragmentally, thus curing all your mental and bodily ailments.' They were as physical as we are, and when their unprepared spacecraft were attacked their crews had been killed, but when forewarned they could form an impregnable barrier around themselves or dematerialize themselves and their craft and move into other dimensions.

As for conditions on their own planet of Aenstria, Shuttlewood was told that medical and scientific knowledge was such that they no longer had any disease, and that they lived many hundreds of years but did not appear to age. They could live on Earth undetected, in their own bases which they had established here and also among Mankind.

Shuttlewood was never sure how genuine these phone calls were. Hoaxing has always been a popular pastime in the UFO world and when there was little activity in the sky, ufologists would sometimes hoax one another just to liven things up a bit. Also there were always the self-righteous sceptics determined to expose gullible UFO investigators. But these lengthy, earnest conversations with strange voices fell into neither of these categories. The Aenstrians had told Shuttlewood that they always used a telephone box on the eastern side of town, known as the Boreham Field kiosk, and they always gave its number when they called him (why they should do this was never explained). One night, during a session of cosmic instruction from Selorik, Shuttlewood thought he would whisper to his wife to ring the police on the other line so that they could race to the Boreham Field phone box and arrest the crank caller. No sooner had the idea entered his mind than Selorik bellowed down the line, 'Shuttlewood, dear friend, do not try and apprehend us. Earth peoples will only suffer if they attempt to do so. We come in peace.'

In 1967, after Shuttlewood had published his first book, which included details of the above phone calls and messages, he received a call from another Aenstrian, named Karne. The Aenstrians had read the book and approved of the publicity given to the messages, but took a dim view of Shuttlewood's imputations of hoaxing. They were also dismayed that he gave the name of the phone box from which they called, for 'this information could alarm those who have habitations in that locality . . . You should not have revealed this, which is of no consequence and hardly secondary importance.'

Quite. In fact it is of no importance at all and one wonders why they made such a fuss about it in the first place, or for that matter, why a technically advanced race needed to use a terrestrial phone. Didn't they know that some of their cosmic colleagues were able to beam a voice signal directly into any telephone line they chose? Karne then continued by lecturing Shuttlewood on the importance of 'trust' and 'truth'. But by then the West Country journalist had had a bellyful of these prating extra-terrestrials. 'Look,' he snarled down the phone, 'if you Aenstrians had the guts and courage to come up to my flat in person, there to be interviewed properly, even if you wear horns on your heads I would have had no hesitation in calling you genuine.' This had no effect on Karne, who replied with a few more homilies to which Shuttlewood replied by slamming down the receiver. Within seconds the doorbell rang: it was Karne in person. He appeared to be a middle-aged man with thick-lensed spectacles, standing 6 feet 2 inches (1.8 metres) tall and wearing a gaberdine mackintosh and brown boots. Declining to shake Shuttlewood's proffered hand, he held his hands together in front of him, fingertips touching and thumbs forming a triangle. This Shuttlewood interpreted as a means of producing a protective circle of energy around the spaceman, who made a short bow and said, 'Greetings to you, Shuttlewood. We bring great love and peace with us. You must learn fully to trust us, before it is too late. Your Earth time is desperately short.' He continued by prophesying some future conflicts, including a third world war, and said that the magnetic stability of the planet was disturbed, which would cause seismic and meteorological disturbances; but 'We were not to worry unduly – the situation was currently being controlled by interplanetary brethren and machines.'

Determined to put this face-to-face interview to some use, Shuttlewood asked if the contact claims and UFO photographs of George Adamski were genuine. Karne replied, 'We are not permitted to give you that information. Adamski was not truly of your cantel [planet] – for the rest, you must look within for the answer.' When the nine-minute interview was over, Karne once more gave a brief bow and

Shuttlewood, in what was intended to be a gesture of friendship, 'firmly grasped his left wrist and thumb', whereupon Karne winced.

Next day, Shuttlewood's sixteen-year-old son, Graham, reported that he had seen the mysterious caller in a local park, and had noted with surprise that he was wearing a bandage on his left wrist and thumb, being at that time unaware of his father's parting gesture, which had apparently resulted in more than just a momentary discomfort. Shuttlewood was in no doubt that he had been contacted by extra-terrestrials, but unlike many contactees he was not wholly convinced of their veracity and purpose. Regardless of the hours of space philosophy and indoctrination which he had undergone, there remained a nagging suspicion that they might be deliberately 'pulling the wool over our eyes for an ulterior purpose'.[6]

Abductee/contactee overlap

The space visitors who met the earlier contactees were, with their perfect physique and handsome blond appearance, idealized versions of humanity. Shuttlewood's visitor with his imperfect eyesight and shabby clothing was much nearer to an average human one might meet on any street. But now we are about to meet the entities who have dominated UFO reports of contact with aliens during the 1970s and 1980s, and the appearance of these creatures bears very little relationship to the average Earth-dweller. The dominant theme of UFO contact in these two decades has been that of abduction. Whereas the contactees who met space brothers were invited into their spacecraft and were treated civilly as guests, abductees are usually first rendered insensible and incapable and then are taken into the spacecraft unwillingly, there to undergo impersonal physical examinations, with little or no communication between themselves and their abductors.

Some UFO investigators believe that there are hundreds or even thousands of people who have had such an experience, very often unknown to themselves, as the abductors also seem to have the ability to erase all memory of the experience. These hidden memories are subsequently brought to light when the abductee becomes prone to irrational fears or repetitive frightening dreams, or becomes aware of an unaccounted-for period of time that occurred immediately after seeing a

UFO, and is persuaded to undergo regressive hypnosis in an attempt to uncover whatever it was that caused the trauma.

Bill Herrmann was one of that small number who emerged from the abduction with a partial memory of the experience. This twenty-seven-year-old car repair manual writer first saw and photographed UFOs over his trailer home in Charleston, South Carolina, in November 1977 and over the next four months saw UFOs on another ten similar occasions. On 18 March 1978 he was outdoors once more watching the enigmatic lights in the sky when the craft swooped down to where he was standing and projected a beam of blue light which drew him into the UFO.

Inside he found the craft was crewed by 3½-foot (1-metre) tall humanoids with hairless grey sponge-like skins and large dark brown eyes set in overlarge heads with under-developed jaws and very small mouths. They wore red one-piece overalls with no visible belts or buttons. He observed these details while he was lying on an examination table beneath equipment which carried five flashing coloured lights. After his tests or examination he was given a tour of the UFO and shown the control room and the power unit room or, as he was told, the 'equilibrium manipulation chamber'. The humanoids spoke to him without moving their lips, which suggests he was receiving the information through telepathy, though he was sure the sound entered through his ears. After a further period of amnesia, he found himself outside on the ground, some 15 miles (24 kilometres) from where he was picked up. He stumbled towards a main road and with the help of the police eventually made his way home.

In May 1979 he was once again taken up the blue light beam into the spaceship and this time the humanoid leader was more communicative as he showed Herrmann around the craft. In the navigation area he was shown a star map and told that they had come from two stars in the Reticulum constellation. Afterwards, while in a trance, Herrmann produced many pages of written material dictated by the aliens. This apparently unique case does not appear to have been examined by experienced investigators, but if genuine it presents a bridge between the contactees and abductees, incorporating as it does elements from the scenarios of both types of report.[7]

Evidence for a hoax?

In 1966 there were still contactees telling of meetings with human-like space people, communicative and friendly. Indrid Cold from the planet Lanulos was one such, and on the evening of 2 November 1966 he landed his 60-foot (18-metre) long spacecraft on the highway in front of Woodrow Derenberger's van near Parkersburg in West Virginia, causing the domestic equipment salesman to come to a stop on the shoulder. Indrid Cold emerged and strolled over to Derenberger. The spaceman was 5 feet 10 or 11 inches (1.7 metres) tall with a tanned skin and dark hair and appeared normal in every way. He stood with his arms folded across his chest, hands tucked under his armpits, and reassured the frightened Derenberger, saying he meant no harm and wished him only happiness. Cold asked about the lights of a nearby city and what Derenberger did for a living, explaining that he was a 'searcher' himself, and was at some pains to reassure the salesman that he was very much like the Earth people. 'Mr Derenberger, look at me. I am the same as you are – I sleep, breathe and bleed even as you do.' All this communication took place without verbal speech, as Derenberger appeared to be able to receive telepathically without difficulty. The conversation lasted some ten minutes, and finally Cold suggested that Derenberger report the meeting to local officials and that he would later confirm the

Woodrow Derenberger claimed contact with Indrid Cold from the planet Lanulos during 1966.

story. With a promise that they would meet again, Cold returned to his spacecraft which lifted up into the night sky.

Later that evening, Derenberger told the police of his experience. They, in turn, told him of two truck drivers who had also been stopped further along the highway by the spacecraft, but these witnesses would not allow their names to be published. During the following twenty-four hours there were several reports in the area of close sightings, although no other witnesses claiming contacts came forward. On the following afternoon, Derenberger spent nearly four hours on television being questioned by a panel of reporters and police.

The following day, 4 November, he was driving home with a friend when Indrid Cold contacted him once more, this time by telepathy only, giving details of his home planet, which he said was very much like Earth, and of his family. His wife was named Kimi, and they had two sons, aged eight and eleven, and a baby girl. The planet was called Lanulos and 'was located close to the galaxy of Ganymede'. This statement appears to be one of the usual nonsensical pieces of information with which the space people sow doubt and confusion, Ganymede not being a galaxy but the largest satellite of Jupiter, which is in our Solar System and, with its ice-covered surface, a most unsuitable place on which to bring up a young family.

If Derenberger had invented the meeting with Indrid Cold to gain fame and attention, then his plot got entirely out of control. In the following months his life and those of his wife and children were completely disrupted. His children suffered taunts and jibes at school and could not sleep at night due to the noise from the crowds of sightseers who continually thronged around their house. As a salesman he found that everyone wanted to talk to him, but only about UFOs, not about buying his goods, and as he lived on his commission his earnings dropped to nothing and savings dwindled alarmingly. To evade the crowds the family moved house four times until they eventually ended up in the anonymity of central Cleveland.

In the following year Derenberger met Indrid Cold several times. Once the spaceman was waiting for him in his back porch when he came home late one night. On another occasion Derenberger was contacted by telepathy as he was driving and Cold directed him to a deserted area where his spaceship had landed. Eventually he was given a trip to the

Amazon, at his own request, and they flew low over the jungle and towns. From there he was taken to the huge 'mother ship' which was parked near the Moon. The usual report of contactees is that the spacecraft they fly in is completely quiet and smooth, with no sensation of motion, but Derenberger found that in this vessel there was 'a continuous vibration' which made him feel slightly nauseated. He was told that it was due to the 'uncontrolled power'. Within the mother ship he met the commander and had a meal of something like potatoes and green beans and a meat that was similar to wild deer. Later there was a flight through the Solar System, where he was able to discover the secret of Saturn's rings. Saturn, it seems, 'is shaped like a big bowl with the people living inside it like in a valley. The rim of this so-called bowl is covered by ice. The rings of Saturn are simply rainbows that are caused by the sun shining on the ice.' Then they flew to Lanulos, coming near to the ground and waving to the people walking there. He was not permitted to land due to the decontamination procedures which would be required.

In May 1967 he was taken on a trip to Lanulos in the mother ship, a journey of some thirty minutes, and he was entertained in Indrid Cold's home. This was not so very different from a modern American luxury home, except for the laundry room, where soiled clothing was hung from hooks and automatic sprays cleaned and rinsed it; no ironing was required. Derenberger then went on a tour of the town with his hosts, noting that he attracted attention because he was the only person wearing clothes. Lanusians were in fact nudists, and they were very curious and somewhat fearful of this strange man from Earth, as they were well aware of the aggressive and warlike nature of the Earth dwellers. Later, however, he took his clothes off and they received him with great friendliness – although he was still noticeable, being, as he says, 'a little bit overweight', whereas all Lanusians had perfect physiques. He was also taken on a trip to Venus where the temperature was always 100°F (40°C) or more, so the Venusians too wore nothing, except 'at formal gatherings or at dances'. They were handsome and healthy and received him with great friendliness. The planet itself had many streams and rivers and lush vegetation, low hills but no mountains, and large fields under cultivation.

Derenberger's early experiences with the curious and the sensation-seekers who thronged his house when he originally announced his contact did not deter him from courting further publicity, and he made various lecture tours and television appearances, all with the object of promoting the space people's philosophy of peace and brotherly love. In March 1968 he was guest on a radio talk and phone-in show where once more he related his Lanulos experiences. One of the phone-in callers was a twenty-one-year-old psychology student who gave the name Ed Bailey and said that he too had been to Lanulos and could confirm what Derenberger had said. This was the first time a contactee had had his story publicly corroborated, and it caused a certain excitement among some UFO enthusiasts and writers on the subject. Ed Bailey generally gave the same details as Derenberger, but said his contact had been called Vadig. Other facts which he said he remembered, Derenberger agreed with and in turn elaborated upon them.

By the winter of 1969 Bailey was a guest on the Long John Nebel radio show, where he was met with the polite scepticism with which Long John usually treated his UFO guests. He was also the lead speaker at a UFO conference held in New York. By this time Ed Bailey, whose name was in fact Thomas F. Monteleone, decided that his student hoax, for that was what his story was, was getting out of hand, and he decided to make no more appearances and to drop out of the UFO scene entirely. He subsequently became a writer of science-fiction stories.

Even among the fantastic reports of many of the contactees, Derenberger's story does stand out. But did he really invent the whole of his reported experiences in order to direct attention to himself regardless of the effect it had on his own life and those of his wife and children? Or was he himself the victim of an even greater hoax, possibly perpetrated by entities whose object was to sow doubt and confusion among humanity?[8]

Spaceships from the Pleiades

The word 'hoax' has been directed at Eduard 'Billy' Meier probably more often than at any other contactee in the history of ufology. Meier is a one-armed unemployed odd-job man living with his Greek wife and three children in the Swiss village of Hinwil, 30 miles (48 kilometres) south-east of Zurich. He claimed that in January 1975 he had been telepathically contacted by a group of space people from the Pleiades who had directed him to visit isolated locations in the surrounding countryside where they landed their 23-foot (7-metre) diameter spacecraft and came out to meet him. To support his story he produced many colour photographs of these spacecraft hovering above the Swiss valleys. These are not the usual blurred, detail-less blobs with which some contactees seek to substantiate their claims, but are sharp and detailed and generally show the valley scenery below the hovering spacecraft. He also produced photographs of Semjase, the space lady from planet Erra who often came out of her landed spacecraft to talk to him.

Over the years Meier has attracted a group of supporters who travel to Hinwil and gather in his kitchen, listening to him relate his experiences and study his photographs and voluminous writings on the Pleiadean culture and philosophy. Critics have pointed to the fact that the Pleiades are relatively young, unformed stars and are neither old

Contactee Eduard Meier seen at his Swiss home; note the portrait of his space friend Semjase.

enough nor stable enough to have formed planets capable of supporting life, but Meier says he was told that the Pleiadeans have emigrated from elsewhere to adapt the planetary climates artificially and form a colony. Meier has amassed many hundreds of photographs and many thousands of pages of written material, and supporters feel that this valuable record should be preserved for posterity, so they have established an institute for that purpose, naming it the Semjase Silver Star Centre.

Meanwhile, some UFO investigators of a more sceptical turn of mind had the photographs subjected to computer analysis. This indicated that the 'spacecraft' were small objects close to the camera: in some instances a suspending line was revealed and in others double exposure and double printing were detected. It was also revealed by UFO investigator Wendelle Stevens that before reproduction in the glossy large-format book *UFO . . . Contact from the Pleiades*, which chronicled some of Meier's adventures, the photographs were retouched and 'enhanced', possibly to remove tell-tale balloons and parachutes in the sky, from which models could have been suspended. Regrettably other interesting photographs of Meier's were not used in this book, such as those he took with the help of his Pleiadean friends when he travelled back in time to photograph the great earthquake of San Francisco as it occurred, and further back to photograph dinosaurs wandering amid the pyramids. Meier seemed to take a perverse delight in seeing how far he could stretch the credulity of his followers, and it seems that with some he hasn't yet reached the limit. He went on to explain that he had also met Jesus while he was in the Holy Land, and had been inducted as the thirteenth disciple. After that, he took a quick tour round the planets photographing the 1975 link-up of American and Russian spacecraft on the way and finally to the edge of the Universe where he managed to snap a picture of God's eye. This has been identified as a fuzzy copy of a picture of the ring nebula in the constellation Lyra. Meier had made his point that he wasn't to be taken seriously. But his photographs of spacecraft flying across the Swiss valleys remain some of the most attractive depictions of UFOs that we have ever seen.[9]

Telepathic messages from space beings

Not all contactees claim to have met people from space. There are many

In 1950 Dan Fry was contacted by spaceman Alan, and went on a trip with him in his craft.

whose conversations with the space people are conducted entirely inside their heads. This need not make the contact any less 'real', for in the final analysis all human awareness occurs 'inside our heads'. In the following chapter we will consider some of these 'channelling' contac-tees, but first let us look at some who straddle both spheres.

Dan Fry is one such: he never met his space contact Alan. In July 1950 Fry was contacted in the desert at White Sands rocket testing

grounds, New Mexico, where he worked as a technician. As he walked in the desert in the cool of the evening, a remote-controlled spacecraft landed near him and the voice of Alan, the space being who was controlling it from a larger vessel 900 miles (1,450 kilometres) above the Earth, warned him to keep clear, speaking in audible tones. After Fry's initial shock, he and Alan carried on a detailed and sometimes technical conversation and then he stepped into the spacecraft for a return trip to New York, the total 4,000 miles (6,500 kilometres) taking about thirty minutes. During this journey Fry was told that, 'The spaceship upon which we live and work and learn has been our only home for generations.' Over the following years Alan contacted Fry twice more, both times by mental means only, and gave him a long and detailed exposition on the cause of and cure for the ills of humanity, urging Fry to use every means he could to spread this message to his fellow men.[10]

In 1951 a Sicilian, Eugenio Siragusa, was waiting for his early morning bus to work when a glowing object appeared overhead and zapped him with a ray of brilliant light. An 'indescribable serenity' flowed through him and an 'inner voice' began to communicate with him, opening his mind to the 'mysteries of Creation' and of his former lives. For the next eleven years these 'Extra-terrestrians', as Siragusa terms them, instructed him, some might say programmed him, by this method. Then one night in 1962 he was prompted to drive to Etna (the volcano in Sicily) and there, near the edge of the volcano, he met two silver-clad figures with long blond hair which hung over their shoulders. One of them spoke in Italian, saying, 'We have been waiting for you, record in your memory what we are going to tell you.'

Then followed a message advocating that humanity should follow true progress and embrace Justice, Freedom, Love and Fraternity. The Brothers of the Inter-Galactic Confederation had travelled through several light-years to warn humanity of the danger towards which it was heading. But instead of receiving thanks, their teachings were ignored or mocked. They assured Siragusa that we were 'supervised by a superior race who will never permit you to come to the disaster of a "nuclear war"'. Siragusa took the space brothers' words to heart and founded a religious cult with headquarters in Geneva, Switzerland, which he named 'Study Centre of the Cosmic Brotherhood'.[11]

The Ambassador of the Elohim

Not only the Brothers of the Inter-Galactic Confederation were concerned about the future of humanity. So too were the Elohim, possibly a rival group of spacemen who claimed that they were the original creators of humanity as recorded in the Book of Genesis. The Hebrew word *Elohim*, meaning 'those who came from the sky', had been mistranslated as 'God the creator', so they claimed. By 1973 they must have decided that events were getting out of control, and it was time to step in, so on 13 December they contacted a twenty-seven-year-old French racing-car journalist named Claude Vorilhon and urged him to visit the extinct volcano of Puy de Lassolas near Clermont-Ferrand.

After enjoying the scenery for a while, he was about to leave when a 'flying saucer' with flashing red lights descended through the mist and landed close to him. The door opened and out came a small being 4 feet (1.2 metres) tall. He was wearing a green one-piece suit and had black hair and beard. Around his head was a 'bubble' of light which shimmered and vibrated, and which Vorilhon likened to a halo. Apart from his diminutive size he appeared almost normal, if one overlooked the greenish tinge to his skin and his elongated eyes. Here, in fact, was an Elohim. After exchanging smiles, Vorilhon asked, 'Where do you come from?' 'From far away,' replied the spaceman. 'Do you speak French?' 'I speak all the languages of the world,' was the reply. 'Do you come from another planet?' 'Yes.'

The spaceman made it clear that this was no chance encounter, but that they had been observing Vorilhon for a long time and had selected him for contact. They went into the spacecraft and sat talking, and the spaceman explained the mission for which Vorilhon had been selected. When he asked for questions, Vorilhon again asked, 'Where do you come from?', but the only answer he was given was 'From a distant planet about which I will tell you nothing for fear that if the men of the Earth weren't wise enough they could come to trouble our peace.' Vorilhon was told to return the next day with notepad and pen and Bible, and the correct interpretation of the sacred writings would be given to him. This he did, and the evocative mythology of the Bible was duly reduced to a mediocre science-fiction adventure of spacemen with atomic disintegrators and anti-gravity beams busily performing the miracles recorded in the Old and New Testaments.

Then Vorilhon's mission was outlined to him. He was to prepare humanity for the return of the Elohim, and the establishment of universal love and cosmic harmony. To do this he was renamed Rael, which was translated as 'The Ambassador of the Elohim', and was advised to collect funds to purchase or build an embassy on Earth. The

174

new commandments which the faithful were to observe were remarkably unvigorous and totally lacking in any moral strictures, being principally concerned with promoting the organization, not overlooking the ten per cent of income which all followers are expected to contribute, and pointing out that 'the Guide of Guides [i.e. Vorilhon] will be the guardian of our embassy, and will be able to live there with his family and the people he has chosen', and also that he would need funds to travel the world, to spread the message. On the plus side the Raelian Movement promotes an activity it calls sensual meditation, which turns out to be nothing other than good old free love. With branches in France, Switzerland, Belgium, Canada and Britain, Claude Vorilhon seems to have tapped an urgent need for a new space-age religion.[12]

The space-inspired Integraton

It is perhaps significant that a UFO cult that originated in France should have as one of its principal tenets the desirability of liberation, whereas one originating in America was more concerned with the retention of youthfulness and the rejuvenation of the old. This was the aim of 'The College of Universal Wisdom', founded by George Van Tassel in the Mojave desert, California, in 1953, after his initial meeting with a man from space. It all started at 2 A.M. on 24 August 1953, when Van Tassel, because of the heat, was asleep outdoors in the desert. He woke to find a man standing at the foot of the bed, and said to him, 'What do you want?' 'My name is Solganda and I would be pleased to show you our craft,' the spaceman replied in perfect English. From then on Van Tassel said nothing: his every thought was intercepted and replied to by the space being before he could form it into a spoken sentence. Nearby hovered a bell-shaped craft, 36 feet (11 metres) in diameter and 19 feet (5.7 metres) high. Stepping into a light-beam shining from beneath the vessel, they floated up into the interior. Van Tassel was given a tour of the spacecraft, seeing the control room, the motor room and the laundry compartment, where their uniforms were cleaned by a 'density of light' process. Solganda accompanied Van Tassel back to his bed and said that he would return some time. He then re-entered his craft, which was out of sight within a few seconds.

The Integratron being built by George Van Tassel, shown on the cover of his newsletter.

⊕ Proceedings

OF THE
COLLEGE OF UNIVERSAL WISDOM
YUCCA VALLEY, CALIFORNIA

A BRANCH OF THE MINISTRY OF UNIVERSAL WISDOM, INC.

A Non-sectarian and Non-profit Organization

for Religious and Scientific Research.

VOLUME 8   JULY – AUGUST – SEPTEMBER 1969   NUMBER 10

At that time Van Tassel was running the Giant Rock Airport near Yucca Valley, California, after a twenty-six-year-long career as a test pilot and safety inspector in the aviation industry. The airport was in

The annual Giant Rock flying saucer convention, photographed in 1966.

fact a remote desert landing strip, but provided Van Tassel and his family with the open-air environment in which they preferred to live. Soon after his visit from Solganda, Van Tassel founded 'The Ministry of Universal Wisdom' and began to contact the space brothers by means of his developing extrasensory perception. Later he utilized their omni-beam by which the space brothers were able to project an audible and visual signal direct into the receiver's brain.

The Giant Rock Airport took its name from a massive lump of granite sitting on the desert floor, some 60 feet (18 metres) high and weighing an estimated 100,000 tons, and it was beneath this that an underground chamber had been constructed where Van Tassel and his associates gathered to sing hymns and space songs and prepare themselves mentally for contact with the space people. Over the years Van Tassel received volumes of space messages comprising cosmic philosophy, advice and technical information, including the design of an Integraton, a dome-shaped building four storeys high which would provide rejuvenation for the aged and prevent ageing in the young. The annual Giant Rock flying saucer convention, which was run by Van Tassel, was for many years a favourite meeting place for contactee buffs and followers of the space brothers. Thousands converged there by car and light aircraft from all over the States, for two days of relaxed meetings in the open air listening to talks from contactees and buying books and pamphlets relating their experiences and the messages of the space brothers.

For twenty-five years Van Tassel laboured to build and equip the Integraton, repeatedly running out of funds and appealing for more money through the pages of his *Proceedings* newsletter. Several times he announced the suspension of the project, but somehow he managed to restart it again, slowly adding to the structure. When he died in 1978, aged sixty-seven, the Integraton was still unfinished, an incongruous structure, gleaming brilliant white out in the desert, and possibly a fitting monument to all the contactees, whose faith in the space brothers has never quite been justified.[13]

12   Space contact through mediumship and channelling

The practice of trance mediumship is as old as Mankind, and traditionally it is a means of contacting unseen entities, usually thought to be the spirits of the recently dead. When in trance, mediums are often controlled by their 'guide', a spirit which acts as a gatekeeper between this world and that of the spirits. In this century some guides have extended their role from that of gatekeeper to one of philosopher and adviser, and in recent years the practice of channelling, as this new form of mediumship is known, has, particularly in the USA, swollen into a growth industry. The mediums we are concerned with here are those who have obtained information regarding life on other worlds, either by speaking to entities from those realms or by direct experience whilst in trance.

Swedenborg and unseen worlds

One of the earliest and most impressive of these seers must be Emanuel Swedenborg, who is today generally known as a theologian and philosopher. He was born in 1688 at Stockholm, Sweden, and after studying at Uppsala University he travelled in Europe accumulating scientific information and becoming fluent in nine languages. He became a metallurgist and mining engineer and made several practical inventions in these areas. He wrote many scientific works dealing with geology, mechanics, geometry, mathematics, physiology and astronomy which were the authorities of their day and were translated into thirty languages. As a relaxation he was a gardener, musician and poet, and in later life was interested in and recorded the symbolism of his dreams.

He had experienced mystical states during his younger years and in 1745 turned his attention to theology and his psychic abilities, which hitherto had remained undeveloped. Various examples of his clairvoyant, precognitive and prophetic abilities are on record, one of the most quoted being the occasion in 1759 when he told the company at dinner that a large fire was at that time blazing 240 miles (385 kilometres) away in Stockholm. Two hours later he announced that the fire had been extinguished, three houses away from his own. Two days later when a messenger from Stockholm arrived, the occurrence was confirmed as accurate in all details. On other occasions he accurately predicted a death and made contact with the dead brother of the queen of Sweden, conveying to her a message of a nature which only she and her dead brother could have known. He said he lived in two worlds at once, and could talk to spirit beings as easily as he could to his fellows.

Our particular interest in Swedenborg concerns his reports of visits to the spirit beings of other worlds, both in the Solar System and beyond, whilst in this alternative state of consciousness. In his work published as *Earths in the Universe* he reports his interaction with the spirits he met when he visited Mercury: they did not use verbal speech but what he calls 'active thought', or as we would now say, telepathy. When Swedenborg expressed an interest in seeing the people of Mercury, he became aware of a man who was 'clothed in a garment of dark blue fitting closely to his body', and the women he saw were beautiful but with faces 'smaller than that of the women of our Earth . . . more slender in body, but of equal height'. They very closely resembled the people of Earth. Their great interest was to collect information relating to ideas, and this they did by examining the memory of any being they met. They also travelled widely and knew of hundreds of thousands of planets which were inhabited.

The spirits of Jupiter Swedenborg perceived as being of a gentle and sweet disposition. Those living in the hot areas of that planet went unclothed and they all enjoyed long meal-times, not to indulge in the food, which was plain and nutritious, but for the pleasure of conversation. Their dwellings were also modest buildings, decorated inside with pale blue cork, and with stars on the ceiling. Swedenborg read to them the seventeenth chapter of St John's Gospel, and they acknowledged that 'all things therein were Divine'. He explained the spread of the Gospels by the use of printing and publishing, at which 'they wondered exceedingly.'

On Venus he found two distinct races, one mild and humane, the other savage and brutal. The mild Venusians were reassuringly Christian in outlook and acknowledged 'our Lord as their one only God'. He was told that the savage beings delighted in robbery, 'and more especially with eating their prey', and decided against taking the scriptures to them. In his travels beyond the Solar System Swedenborg visited several 'Earths in the Starry Heaven', most of whose inhabitants seemed to live a simple agrarian life. On one planet he found that the inhabitants constructed their temples from the living trees, by training and pruning the limbs, thus forming the floor, walls, roof and stairs from the living trunks and boughs. Here too they were surprised to learn that on Earth information of all kinds was spread by the printed word, but conceded that the divine revelation and instruction from angels which they enjoyed would be unsuitable for inhabitants of Earth.

After twenty-seven years of interacting with unseen worlds, Swedenborg foretold the day of his death, and on 29 March 1772 he duly died, leaving some of the most enigmatic writings ever penned.[1]

The Spirits' Book by Allan Kardec is a fascinating collection of ideas and possibilities expounding on the Immortality of the Soul, the Nature of Spirits and their Relations with Men, Present and Future Life, and the Destiny of the Human Race. The author was Léon Rivail, born in 1804 at Lyons, France, and as a Protestant in a mostly Catholic country he was the victim of religious intolerance in his early years. This helped to turn his thoughts to the need for and value of education, and he made his life's work in this field. He ran a school for boys from 1828, and in 1830 he started to give public lectures on the sciences, which were always well attended. He also wrote several books dealing with educational matters.

During the mid-nineteenth century the craze for mediumistic table-rapping was sweeping through the Western world, and Rivail's enquiring mind was soon investigating the phenomenon with the help of two young women mediums, daughters of a friend. Their customary trivial communications acquired a more serious note when he was present, and he was soon communicating, via the mediums, apparently with spirits of superior intellect and intent. These communications completely revolutionized his ideas and convictions and he determined to collect and publish the teachings on the new philosophy.

It was at this time that the communicators gave him the title for his book, *The Spirits' Book*, and told him to use the pseudonym Allan Kardec. The book was in fact compiled over a period of fifteen years from the work of many different mediums, none of whom are identified and so we do not know who received the following answer to the question 'Are all the globes that revolve in space inhabited?' The answer was: 'Yes; and the people of the earth are far from being, as you suppose, the first in intelligence, goodness and general development. There are many men having a high opinion of themselves who even imagine that your little globe alone, of all the countless myriads of globes around you, has the privilege of being inhabited by reasoning beings. They fancy that God has created the Universe only for them. Insensate vanity!' The questioner went on to establish that planets differ one from another physically as did the beings on them, and those distant from a sun were warmed and lit by other means, such as electricity.[2]

A more direct means of obtaining interplanetary information was employed by Sherman Denton in 1866. He claimed he could transport his senses to other planets and see and speak with the people there. In fact his whole family claimed to receive messages from space through psychic vibrations, but Sherman was the best, or the boldest. In his public lectures he explained that the Martians could easily pass as Earth people, and were here in great numbers to gather knowledge about Earthmen. At home, on the Red Planet, they lived in an ideal society, and used flying machines made from aluminium to travel on their own planet and across space.[3]

Writing the Martian language

Even more fantastic were the revelations of Swiss medium Catherine Elise Müller of Geneva, who in the 1890s described her visits to Mars. She drew pictures of the Martians she met, the plants and animals she saw, and the houses in which the people lived. She also spoke and wrote the Martian language, and this was of particular interest to Theodore Flournoy, Professor of Psychology at Geneva University, who as a psychic researcher attended her seances. At that time the possibility of life on other planets was being avidly discussed. Some astronomers

Martian writing as presented by medium 'Hélène Smith' in 1898.

were identifying canals on Mars and speculation was fuelled by the science fiction of Flammarion and Verne. But Hélène Smith, the pseudonym used by Flournoy when he published his case study, was limited to the knowledge immediately available to her subconscious. The Martians she met displayed no advanced technology or high moral philosophy. Her drawings of the Martian architecture and landscape were evidently influenced by the designs on the Far Eastern porcelain dishes and lacquered furniture that were in great vogue at the time, and the Martian dress and people had a homely peasant-like appearance.

Of more interest to Flournoy was the Martian language which Müller spoke and wrote fluently. It was, he found, a genuine language, with its own coherent grammar and construction, able to express ideas in a continuous and logical sequence. By means of detailed analysis of the syntax and the numerical frequency and musical pitch of vowel sounds, he was able to satisfy himself that the language was derived from Müller's native French, but none the less was a brilliant example of subconscious creation.

In passing we might add that Müller's mediumistic creations were not limited to her Martian cycle. Earlier she had been a reincarnation of Marie Antoinette, and at another time the Arab wife of a fifteenth-century southern Indian prince, these also being genuine but extremely obscure historical figures, information on whom it would have been difficult for her to acquire consciously. She had at times also produced apports of foreign coins and flowers out of season, and done religious paintings whilst in trance. In all, Catherine Müller's mediumship provided a rich field for psychic researchers for many years.[4]

Harry Price, the British psychic researcher, arranged in 1926 to sit with medium Mrs St John James, who believed she was in communication with various Martians. When in trance, she spoke in a strange tongue, sang 'unearthly' songs, 'rather like a solo by a crowing cock', as Price later wrote, and filled pages with Martian writing and sketches of the Martians. Price concluded: 'I am afraid that mediums have told me very little about Mars and the Martians.'[5]

For many years an American, William Ferguson, had practised a relaxation method which would enable him to raise his molecular vibrations and enter higher dimensions. In July 1938 he succeeded in visiting the seventh dimension and returned with the ability to space-travel using his fourth-dimensional body. In 1947 he was drawn up to the planet Mars where Khauga, Celestial Scientist, was waiting to question him and show him around. Their beautiful cities, which were isolated from the planetary atmosphere by force fields, were constructed on a linear principle and had been mistaken by astronomers for canals. But within the protective envelope they maintained an atmosphere of 76°F (25°C), which enabled them to control the weather and grow vegetation continuously. In the Central Boulevard, which was a combination of city square, parkland and conservatory, he met the welcoming committee of 'Uniphysicists', and observed that everyone was a redhead. They were on average a foot (30 centimetres) shorter than Earthmen, wore cream and gold robes, and with their shoulder-length ginger locks presented a most striking appearance.

Ferguson was, Khauga said, 'the first earth man we have ever been able to bring to our planet'. Being 20,000 years in advance of the Earth in all areas of development, they were going to help us on to the next evolutionary step, and would be releasing large quantities of expanded elements over our planet to speed events up. They would also release positive energy to counteract the negative energy which Mankind was producing. After two hours on Mars, Ferguson was returned to Earth but had some difficulty in being seen by his family, until he lay down for a while and let his molecules slow down to the third-dimensional vibration.[6]

Co-operating with the space people

Paul Solem of Idaho never claimed to have left this planet, either in his body or out of it, but he said he was in telepathic contact with Venusians in their flying saucers and to prove it could call them to appear overhead. This first happened in 1948 when he received the mental message, 'We are from another planet. You will hear from us later' as a formation of three glowing discs sped overhead. Four years later he thought he had met a Venusian 'angel', as the space being identified itself, and was told that his task was to work with the American Indians and prepare for the fulfilment of their prophecy regarding the coming of the Day of Purification, and the establishment of a New Age on Earth.

By 1969 he was addressing Indian meetings, foretelling the imminent arrival of the great Day of Purification when the true and the faithful would be transported by spacecraft to safer planets while the laggards would perish on Earth. He established himself in Prescott, Arizona, traditionally a centre for contactee enthusiasts, where he demonstrated to local reporters his ability to contact the space people and bring them flying overhead. The editor of the *Prescott Courier* watched as a light moved about the sky in an inexplicable manner, and later wrote, 'A flying saucer? Yes, if we could believe our eyes.'

In 1971 Solem announced a major demonstration in daylight when a flying saucer would come down within a few hundred feet. Over one thousand hopeful witnesses arrived and the media were well represented; the only thing missing was the brothers in their spacecraft, who completely failed to make an appearance. Solem, having staked all on this ultimate demonstration, had no option but to fade from the scene as soon as possible, and since then has disappeared from ufology.[7]

At the 1960 Flying Saucer Convention at Giant Rock in the Mojave Desert of California, a wild-eyed figure spent the morning accosting the attendees and warning them to expect something strange to happen that afternoon. He was Allen Noonan, outdoor signwriter and contactee, who had just received a mental message from the United Planetary Organization. The Sun was burning down on to the convention area by the Giant Rock in the treeless desert and the message was that at 2.30 P.M. the UPO 'would furnish shade'. A little before the appointed time Noonan received a second message telling him to concentrate his mind on the Sun. This would form a beam upon which the spacecraft could home in. In Noonan's words: 'Then all of a sudden, clouds began to form between us and the sun. The crowd just went wild.'

He demonstrated this co-operation from the space people again in 1967 when journalist Lloyd Mallan interviewed him in a hotel in Los Angeles. Going to the window, Noonan concentrated and soon had two small clouds floating across the clear evening sky. Mallan felt a chill go up his spine but captured the eerie sight with his camera, not completely convinced by Noonan's explanation, but certainly left wondering.

The ability to affect cloud formations was also demonstrated in 1954 by Dr Rolf Alexander at Orillia, Ontario, when over a period of eight minutes he disintegrated a cumulus clump, selected from a largely cloudy sky. The doctor, whose MD was quite genuine and awarded at Prague, had also been a pupil of Gurdjieff and had lived in India and Tibet studying Yoga and Buddhism, as well as Huana in the Polynesian Islands. He had in fact sought out knowledge of the most effective mind control disciplines from both West and East, and had melded them into a system he called 'Creative Realism', which when applied by any individual would lead to a more complete and fulfilling life. Cloud-busting was a mere offshoot of his mental control.

Possibly a similar form of controlled psychic energy was, unwittingly, used by Noonan with his cloud creations, but as with other contactees his initial plunge into the psychic realms had been so emotionally overwhelming that he had jettisoned all sense of proportion. His initial experience occurred while he was at work one day, painting an outdoors advertisement. Involuntarily his consciousness left his body and he was standing before a great white throne in a building of light. Out of the brilliant white light a booming voice asked, 'Will you agree to be the Saviour of the world?' Noonan accepted the post and thenceforth proclaimed himself to be a Cosmic Master and the New Messiah, which effectively turned most people off, notwithstanding his cloud control demonstrations.[8]

Interplanetary intermediaries

In May 1954 another Cosmic Master was established by the audible
command 'Prepare yourself! You are to become the voice of Interplanetary Parliament.' George King, variously described as a chauffeur
or taxi-cab driver, and a long-time student of Eastern philosophy and
spiritual healing, was tidying up one Saturday morning in his London
bachelor flat when the voice, of a 'peculiar musical and deep penetrating
quality', spoke in the room, without any presence being seen. After
brushing up his yogic trance techniques, King was soon channelling the
words of wisdom from the Venusian master Aetherius to an enthralled
crowd in London's Caxton Hall. Other communications were received
from entities such as Mars Sector 6 and Jupiter Sector 92, who extolled
the peace and beauty of life on their planets. They also revealed details
of how their lives were organized. The Martian canals, for example,
were in fact strips of vegetation planted along the lines of magnetic force
which crossed the planet, and the smaller of the Martian moons was an
artificial satellite used to control the weather conditions on the planet.
King had himself on one occasion visited the Red Planet but had had
little time to note any interesting details as he was occupied with
assisting the Martians to overcome a particularly nasty Android which
had arrived in Mars' orbit riding on a meteor.

His visit to Venus was far more relaxed. Travelling once more in his
etheric body, he landed in the Valley of the Sun at the Temple of
Solace. His guide, Patana, took him on a tour of the valley where he saw
another temple, the dome of which emitted a blinding white light. Here
was the spiritual centre of Venus from which radiations penetrated into
every Venusian mind. King then experienced 'a supreme, pulsating,
scintillating, living brilliance which knew me more completely than I
had ever known myself.' Soon he was back in his London room and set
about organizing the Aetherius Society, first in London and later in Los
Angeles where its headquarters now are. Then in 1958 he was the
channel for the great Venusian master, Jesus, who delivered to Mankind his new cosmic teachings of 'The Twelve Blessings'. Not long
after, 'Satellite No 3' began to orbit the Earth. This huge Martian
spaceship periodically releases spiritual vibrations which the Aetherius
Society members redirect to bring about peace and enlightenment

among humanity. Although not claiming to be a new messiah, King has always enjoyed the distinction of a title and soon became the Reverend DD. His most recent elevation was to the royal heights of 'His Serene Highness, Prince George King de Santorini', his coronation being held at St George's church in Mayfair, London.[9]

Possibly no one gets more fun from being a channeller to the space people than Ruth Norman, who with her husband Ernest founded the Unarius Academy of Science in 1956 and opened the Center for the New World Teaching at El Cajón, California, in 1975. Variously known by her disciples as the 'Universal Seeress', 'Archangel Uriel' and 'Ioshanna', she receives cheers and thumbs-up signs when she rides down the California freeway in her 'Cosmic Car', a large automobile with its sides decorated with paintings of spacecraft and sporting a large model of a flying saucer complete with flashing lights on its roof. Known to the locals as 'Spaceship Ruthie', Norman evidently believes that nothing succeeds like excess, and she regularly appears with elaborate hairstyles and wearing diaphanous flowing gowns of brilliant colours studded with sequins. She is in communication with the other thirty-two 'Planets of the Interplanetary Confederation', of which Earth is the thirty-third. 800,000 years ago the Confederation was a successful organization with interplanetary travel between all planets, but after a devastating interplanetary war communication ceased and has only gradually been re-established in recent centuries. The living conditions on the other thirty-two planets are known, but out of consideration to our readers we will mention only two. Planet Yessu is led by a male named Mang, the people are brown-skinned and live in feudal conditions. War was rife for many years until Mang made contact with 'Archangel Uriel' who persuaded him to make a truce between the nations, whereas on Planet Valneza, whose leader is a lady named Dera, the people are peaceful and worship nature spirits. They have no science, live a simple agrarian life, and have a very limited concept of spiritual evolution. Ruth Norman may not be very convincing as an interplanetary intermediary, but there is no doubt she has brought a lot of fun into the world of channelling.[10]

Of a more serious mien is Michael Blake Read, an Englishman who has lived and worked in Canada for many years. In the early 1970s he found he had mediumistic abilities and has since had thousands of channelling sessions throughout Canada and the United States. With his associate, Philippa Lee, acting as trance-session director, they have received numerous communications from a group of discarnate entities whom they have named the 'Evergreens', who say they are between existences, waiting to reincarnate but separated from humanity by a different rate of vibration. They have supplied much information on the entities from the star system Boötes who are bisexual, bifurcated and humanoid. Their planet is darker and hotter than Earth, with less humidity and a high level of methane in its atmosphere. A description of their culture would be almost meaningless to us; they do, for instance, gain intense satisfaction by gazing at an area of yellow colour for many minutes. Our ability and pleasure in reading a printed page is to them incomprehensible, and yet they can understand and speak many Earth languages by having listened to radio broadcasts. For many centuries they have periodically visited Earth as well as many other planets, and had given the Atlantean civilization a knowledge of mathematics.[11]

Mental messages were also received by Luis Muzio Ambrosio, a Brazilian medium, in 1967, but in his case they were beamed from a hovering spaceship. The craft responded to his flashed headlamps, but when he moved to leave the car a pencil-thin beam of brilliant green light hit his chest and travelled up to his head, and all mental and physical volition left him. Within his head he heard a voice which said, 'We are friends and we come from the planet Venus . . . Do not be afraid . . . We are on a peaceful mission and desire the well-being of the people of Earth . . . Be calm . . .' As a medium accustomed to communicating with disembodied entities, this experience may not have had the same shock value for Ambrosio that it might have produced in others. He had in fact received a similar mental message in 1964, when in bed one night he saw an orange globe in his room and heard a voice saying, 'We are friends and we are in our spaceship on the roof of your house.' Later neighbours confirmed that a glowing object had been hovering above Ambrosio's house.[12]

Star people incarnate on Earth

Is a hovering Venusian spaceship more, or less, believable than a claimed ability to leave one's body on Earth while the consciousness travels to other planets? This has happened a number of times to William Goodlett of Salem, Virginia, who in 1968 had his first involuntary experience. For a few brief moments he found himself viewing a strange rocky landscape with a small pale sun on the horizon from within a thin, green-skinned, 4-foot (1.2-metre) tall body. He thought he could recognize the Earth and Jupiter in the sky, and so was probably in the Solar System, on an orbiting body between the two. Another time he found himself on a planet inhabited by beings very much like humans, again only 4 feet (1.2 metres) tall with large eyes and long hair. They lived in rock-hewn dwellings along the side of a large canyon; of particular interest was their atmosphere, which was as dense as a thin liquid. As Goodlett moved along the street, observing the phosphorescent glow which illuminated the interiors of the houses, the inhabitants gave him big grins; they were friendly and not unused to strangers appearing in their midst. During the following years Goodlett visited several other planets in this disembodied state, each planet having its own strange character but none being totally alien or inimical.[13]

Goodlett, we are told, is one of that group of humans who are 'star people', that is entities from other planets who have chosen to incarnate on Earth to assist humanity in its development, although as Earth dwellers they are not usually aware of their origins, history or present purpose. This contrasts with the 'spacemen living on Earth' who are described in the following chapter, for the latter retain their extra-terrestrial form and their non-earthly personality and memories throughout their sojourn on Earth.

One might expect that the entities who choose to 'become' Earth people would arrange to be people of some influence or power, but this does not always seem to be the case. In rural Colorado, there lives Joyce Updike, book-keeper, housewife, and mother of seven, who for many years was puzzled by her long history of half-remembered dream-like contact and abductee memories. Finally she consulted Dr Leo Sprinkle, psychology professor at the University of Wyoming, who has achieved some renown with his annual conference where contactees and abductees gather to exchange experiences and receive mutual support. Under hypnosis, her various contact experiences were recalled, and later when she consulted Ruth Montgomery, author and channel for spiritual advisors, she learned that she was a highly developed being from Sirius who had replaced the original occupant of the Updike body, by agreement, in order to experience Earth life with all its joys and sorrows, such an entity being known as a 'Star Walk-in'.[14]

Saving the world

Many channels worldwide receive messages from 'Ashtar', sometimes known as the 'Ashtar Command', covering a wide range of topics. These range from space philosophy and advice on life conduct, to astronomical information and warnings of impending cataclysms with plans for world evacuation. 'Ashtar Command' is in a fleet of spacecraft which have been in space orbit for millennia and so they should be able to impart accurate information on other planetary conditions. They say that there are many 'wondrous stars which are really suns! Going round them in orbits greater and smaller are many planets where there dwell beings often as you, yourselves. And these are only in your own universe! Beyond these are many universes greater and more wonderful than that which is your own!' They monitor Earth's condition constantly and for several decades have predicted a shift in the Earth's axis to be imminent; they wait ready to put the Great Evacuation into operation. As the shift occurs small spacecraft will levitate 'earthlings whose superior development will be required in the New Age' up to the huge mother ships orbiting thousands of miles above Earth, to be returned to Earth once events have settled down and the New Age can start.[15]

In 1974 Dr Andrija Puharich, American scientist and psychic investigator (whom we met working with Uri Geller in Chapter 9), was working at Ossining, New York, with a small group of colleagues, among whom was trance medium Phyllis Schlemmer. They were regularly in communication with a group of disembodied entities known as 'The Nine', who were 'pure light beings', existing 'at a velocity beyond light, beyond photons, beyond tachyons'. Hundreds of hours of communications were taped and many fascinating and, if accurate, important matters were conveyed to the group. Whereas the Ashtar Command predicted an axial Earth shift and promised evacuation of selected humans, The Nine foretell a massed landing on Earth of many different types of spacecraft. The space people would stay to help

Mankind with its problems by introducing new technologies, but they would also come to raise the level of spiritual awareness throughout the planet.

This theme of the Earth people's undeveloped spirituality is a recurring one in the messages from The Nine. In order to convey to the group the full significance of this theme, The Nine arranged for Phyllis to make an out-of-the-body journey to their dimension, where they instructed her in greater detail. On her return she was exuberant with the new knowledge she had been given by The Nine, who were a 'cosmic governing council', and who explained that because Earth people were lagging behind in spiritual development this was holding back the development of the Solar System, the galaxy and ultimately the master plan for the Universe itself. Because every intelligence within the Universe had to incarnate upon Earth at some time, planet Earth was an essential link in the ultimate scheme, and could not be neglected any longer. But many beings had become enmeshed in material desires and having lost sight of any spiritual attainments were reincarnating upon Earth time after time. Puharich and his group were also told that they had reincarnated on Earth several times, and had chosen to do so in order to gain an understanding of life on the planet and so be able to raise its level of consciousness. They were in fact acting as channels for energy from The Nine for this purpose.[16]

This is a regular feature of contactee literature. The space entity is always at pains to assure the contactee or channeller of their unique value and importance in bringing enlightenment to Mankind and saving the world. Solem, Noonan and King are all convinced of their unique mission and dedicate themselves to serving their contacts, though these never seem to have any connection or resemblance one to another. Somewhat more consistency is displayed by the receivers of communications from Ashtar, of which there are many dotted all over the globe. Assuming that these channellers are not consciously copying each other and using a popular name for their pronouncements, the Ashtar communiqués from different sources do bear a certain resemblance to one another. But they fail to present a convincingly elevated philosophy for humanity to adopt. Apart from the repeated assurance that the favoured few will be airlifted to safety, which is, in essence, a standard element in most traditional religions (but why spiritual beings with a philosophy of the continuation of life beyond death should feel the need to preserve the physical body is never satisfactorily explained), there is also much woolly proselytizing about living with cosmic love and harmony and raising Earth's spiritual vibrations, without a single suggestion on the more pressing need for individuals to increase their self-awareness and improve their relationship with the planet on which they exist.

Much of the channelled material appears to emanate from the channeller's subconscious, and largely consists of reassuring platitudes designed to induce a feeling of well-being in the listener. This appeals to those who are spiritually inclined but mentally indolent; such people tend to feel that listening to homilies is virtually as good as acting on them. Perhaps the communications which are more likely to have come from a non-human intelligence are those which are least interesting or intelligible. A non-human entity might well have difficulty in communicating unimaginable concepts to Earth-dwellers, and would very likely use language in an incongruous, non-idiomatic or even unintelligible manner. We should not expect literature or poetic expression from an alien mind, but certain themes do seem to be frequently voiced, and the possible significance of these will be considered in Chapter 14.

13   Spacemen living on Earth

If we are to believe what the extra-terrestrials tell us, there are many thousands of spacemen on Earth, living and working with humans, but unrecognized for who they really are. The 'man from another planet' who visited Dr Leopoldo Diaz in 1977 (an event to be described more fully later in this chapter) told him that there are 10,000 spacemen dwelling among us, living and working in all walks of life, but here to help us and 'not intervening in any way'.[1] When Peter, who had a strange UFO encounter in South Africa (see Chapter 8), was hypnotized in 1974, he revealed that thousands of spacemen are among us, working as 'clerks, typists, businessmen, university students, lecturers, dustbin cleaners . . .', but they never do anything that would reveal who they really are, and work by influencing people to do whatever it is they want.[2] The three Venezuelan men who spoke to entities from a landed UFO in 1965 (see Chapter 7) were told in answer to their question that there were precisely 2,417,805 spacemen living among humans on Earth.[3]

Contactees meet extra-terrestrials living on Earth

The famous contactee George Adamski was also told that many space people are living undetected on Earth. He wrote about his many casual encounters with them in his book *Flying Saucers Farewell*:

> Through the years since I first met Orthon on a California desert, I have had many meetings with our space-travelling friends. Some have been very casual and unexpected. Others I expected, much as I described in *Inside the Space Ships*. Never have I been able to make definite appointments with them for a specific time and place, nor

have I outgrown the inner exaltation of being in their company. However, even though I have had so many meetings with them, it would be as foolish for me to say that I know all who are on Earth, as it would be to say that I know everybody in any city or town. I have been told that on many occasions I have been visited by and talked with space travellers without recognizing them, and without their identifying themselves. On a few occasions, I have later met one or two on a ship whom I recognized as having talked with previously, without recognition.

It is for this reason that I have so often said and written that many people, in fact untold numbers of people, have met and talked with space travellers without recognizing them. Many work in industries and government positions throughout the world. They may also be found in the armed forces of every nation, working in divisions of science, communications, medical corps, etc. where they are not required to be trained for slaughter of their fellow men.[4]

Adamski also described a specific encounter with a spaceman on Earth in *Inside the Space Ships*. It was on 18 February 1953 when he was in Los Angeles. During the evening he felt that something was going to happen, but by 10.30 nothing had occurred and he was disappointed. Then two men looking like young businessmen approached him, and one addressed him by name. He realized who they were, and went out with them, travelling in their black Pontiac car out of the city. They told him they were from Mars and Saturn, and the Martian said,

We are what you on Earth might call 'Contact men'. We live and work here, because, as you know, it is necessary on Earth to earn money with which to buy clothing, food and the many things that people must have. We have lived on your planet now for several years. At first we did have a slight accent [in answer to Adamski's unspoken thought about their good knowledge of English]. But that has been overcome and, as you can see, we are unrecognized as other than Earth men.

At our work and in our leisure time we mingle with people here on Earth, never betraying the secret that we are inhabitants of other worlds. That would be dangerous, as you well know. We understand you people better than most of you know yourselves and can plainly see the reasons for many of the unhappy conditions that surround you.

We are aware that you yourself have faced ridicule and criticism because of your persistence in proclaiming the reality of human life on other planets, which your scientists say are incapable of maintaining life. So you can well imagine what would happen to us if we so much as hinted that our *homes* are on other planets! If we stated the simple truth – that we have come to your Earth to work and to learn, just as some of you go to other nations to live and to study – we would be labelled insane.

We are permitted to make brief visits to our home planets. Just as you long for a change of scene or to see old friends, so it is with us. It is necessary, of course, to arrange such absences during official holidays, or even over a week-end, so that we will not be missed by our associates here on Earth.

After a long journey they turned off the highway and arrived at a 'soft-white glowing object' on the ground. Beside the spacecraft stood another man whom Adamski recognized – his friend Orthon from Venus. The three took Adamski up into the ship, and he was given a conducted tour, as well as a flight out into space and a landing on a 'mother ship' 2,000 feet (600 metres) long.[5]

Other contactees also saw spacemen on earth – or thought they did. Truman Bethurum, whose experiences with the space people are described in Chapter 11, thought he saw Captain Aura Rhanes in a restaurant at Glendale, Nevada. She was from the planet Clarion, and during 1952 Bethurum met her eleven times. When he saw her in the restaurant, he spoke to her but she denied knowing him. When she and her companion left, the waitress brought him a message: 'The lady told me to tell you that she knows you and that she was sorry and yes is the answer to some of your questions.' A friend waiting outside told Bethurum that no one had gone out through the restaurant door.[6] Orfeo Angelucci (see Chapter 10) saw his friend from space, whom he called Neptune, in the bus terminal at Burbank, California, one evening in October 1952. He looked like an ordinary person in a dark business suit, a dark felt hat and carrying a briefcase, but Angelucci was sure it was

Neptune. However he looked stern, not smiling, and Angelucci received a strong telepathic command not to approach him. As Angelucci stood at the news-stand thumbing through a magazine, Neptune sent him a telepathic message:

> The last time you saw me, Orfeo, I was in a less objectified projection in your three-dimensional world. The purpose being to give you some idea of our true aspect. But now tonight you see me fully objectified. If you did not know who I am, you could not tell me from one of your fellows. Tonight I am no half-phantom, but can move among men as an Earthman. It is not necessary for you to speak to me; you have gained the understanding. You know now that we can appear and function as human beings.[7]

In November 1968 contactee Lester Rosas was on the beach near his university in Puerto Rico when a man with long blond hair approached and gave the password that Rosas had received from his space friend Al-Deena (see Chapter 10 for more details of his space contacts). The man identified himself as Vi-Dal from Venus, and said he was the same spaceman whom George Adamski knew as Orthon. He was here 'to help my brothers of other planets in their missions here on your beautiful island . . . We are "keeping tabs", as you say, on what the Arecibo Observatory is doing regarding space exploration' (some of the work of the Arecibo Observatory was described in Chapter 3). He wore his hair long so as not to be conspicuous: 'I won't be spotted because I just look like a hippie, and it would be the last thing anyone would think of that I am not of this Earth!' They then talked at great length for two hours, and one of the things Vi-Dal told Rosas was that George Adamski is now reincarnated on Venus.[8]

Vi-Dal also mentioned contactee Howard Menger (see Chapter 10), whose involvement with spacemen on Earth was rather more active than that of other contactees, for during the 1950s he was involved in helping to integrate spacemen into human society. He bought clothes and took them to the points of contact, but he had difficulties with the female garments. The bras he took them were rejected with giggles – they couldn't wear them, and never had. The high heels too took some getting used to, and they complained, 'Why can't your women wear sensible shoes?' Menger had to cut the hair of a spaceman newly arrived on Earth: it hung to his shoulders and would have been immediately noticeable in those days of very short haircuts. He also briefed them on 'customs, slang and habits', and took them the food they asked for. This was frozen fruit juices, tinned fruit and vegetables, wholewheat bread, wheatgerm, etc. They refused milk, and also fresh citrus fruits. They never asked for identification papers or help in getting jobs, and presumably had no difficulties with things like this once they had made the initial adjustment to Earth life.[9]

It is rare for the spaceman himself to write a description of his visit to Earth, but one who did so was Valiant Thor, who allegedly came from Venus. Over seventy of his fellow Venusians were also working here, along with Earth people, and his 1959 landing near Alexandria, Virginia, was described in his own words in Dr Frank E. Stranges' book

A photograph said to show Venusian Valiant Thor (right), while he was living on Earth.

The Stranger at the Pentagon. Val followed the precept that sceptics always insist upon: he sought to meet the leader of the 'Free World'. Amazingly, he was met from the spaceship by a police patrol car and taken straight to the Pentagon for a meeting with the President![10]

Not all the spacemen on Earth came down directly from spacecraft. Some are said to have gone to the UFO bases at various sites on Earth and under the sea before going out to live among the people. Mention has already been made of various terrestrial UFO bases, but one early contactee, Guy Ballard, published details of his encounters twenty years before the main stream of such reports in the 1950s. In 1935 he told how he had met a mysterious being on Mount Shasta in California, and had also joined in a gathering of twelve Venusians held inside Royal Teton Mountain. The meeting-place was a vast cavern inside the mountain, and the Venusians played the harp and violin, as well as showing scenes from Venus on a large mirror or screen.[11]

It is not very clear what the spacemen on Earth are up to, especially the ones who claim to be living and working unrecognized among Earth people. Are they simply enjoying a change of scene, or do they have some more sinister purpose? Those who believe that extra-terrestrials are evil and intent upon destroying Earth see them as working quietly to undermine our civilization; while those who believe them to be benevolent see them as a silent force for good. It is quite possible to become paranoid about this, and to blame any undesirable developments on the influence of Earth-based extra-terrestrials. However it seems to us that Mankind is quite capable of engineering its own destruction without having to implicate the (probably perfectly innocent) extra-terrestrials; indeed they could be working overtime to try and save us from ourselves. We may never know, for they do not seem keen to publicize their activities.

Contact with Ummo?

Even those extra-terrestrials who do reveal their presence to Earth people do not convey very much information about themselves and their planet, but tend to speak vaguely without giving hard facts, and certainly without conveying any scientific information that is not already known on Earth. This does not apply to the Ummites, however, whose activities in the 1960s and 1970s caused bafflement among ufologists, especially in Spain where the majority of the contacts took place. It all began in 1950, when some people from Ummo, a planet 14.6 light-years away and possibly circling the star Wolf 424, landed on Earth and infiltrated into human society. In 1965 they began to make contact with certain people, mainly but not exclusively in Spain, to explain what they were doing on Earth. Some contacts were made by telephone, others by letter, but invariably complex technical information was conveyed. In 1967 a UFO landing was promised, which duly occurred and photographs were taken by witnesses – but when analysed by computer definite evidence of a hoax was found, and the 'UFO' appeared to be a paper plate or something similar suspended by a thread.

Yet despite this, the Ummo communications remain tantalizing. An enormous amount of technical data was being passed, and the Ummites were using eight different Earth languages. There were claimed to be eighty Ummites living on Earth, the team members changing from time to time. They claimed that their 'mission on Earth' is for analysis and study, as they wish to learn 'the geographic structure, orogenic movements, biosphere, human social structure and its history and evolution, conditions of the atmosphere and structures of the principal planet and of the other planets, including one located beyond Pluto'. The communicator continued, 'Our forms of thought prohibit us from interfering in your social evolution. Any contribution from our culture to yours transcending simple information about our science and customs would naturally imply an interference in that evolution which, from our deontological perspective, would be damaging.'

Two volumes of communications, covering a wide range of subjects, have been published by the Spanish ufologist Antonio Ribera, one of the people contacted by the Ummites, as *UFO Contact from Planet Ummo*.[12] If this case really was a hoax, whoever perpetrated it must have been working full time to produce all the material; if others were involved, it is strange that no one has yet given the secret away. If it was genuine, it is strange that no truly incontrovertible evidence for the existence of the Ummites resulted from the mass of material conveyed.

Spacemen medically examined

The corpse of a dead extra-terrestrial would no doubt help to persuade people that they really do exist. There are rumours that the US authorities already hold such corpses in cold storage, but rumours they are likely to remain. On at least two occasions extra-terrestrials have allegedly offered themselves for physical examination by terrestrial doctors (which makes a welcome change from the abduction of humans into spacecraft and their medical examination). This happened in Caracas, Venezuela, on 7 August 1967, when a 4-foot (1.2-metre) being entered the office of Dr Sanchez Vegas and asked for an examination. He told the doctor not to worry about his high temperature, as he did not come from Earth. He claimed to have learned his perfect Spanish through the use of a machine; but he did not understand Dr Vegas when he asked the entity his age, and told him also that their method of reproduction was different from ours. The doctor discovered some strange physical characteristics: the being had no ear holes, his eyes were round, his heart made noises like a human foetus, and he had ten teeth, five above and five below, a double one in the middle and two either side. He allegedly told the doctor that life on his planet was more advanced than on Earth, with diseases and wars eradicated, and that he was here to take some scientists back to his world so that they could become up-to-date and thus help progress on Earth. This contrasts with most other spacemen's avowed refusal to help Earth by giving scientific knowledge and thus interfere with our evolution. At least four witnesses claimed to have seen a UFO parked in front of the doctor's building that same day, and one man said he had seen a small entity which flew out through his open bedroom window at 2 A.M. when the witness jumped out of bed. Outside were bright multi-coloured lights which dazzled him. Later he found marks outside, including footprints.[13]

Nearly ten years later, a doctor in Guadalajara, Mexico, was also asked for a physical examination by a man from another planet. On 28 October 1976 Dr Leopoldo Diaz found the man waiting in his office. When examined as requested, he was found to be totally normal and healthy, in fact healthier than many men. He then revealed he was not human, but from a planet beyond the Sun. Although physically very similar to humans, he was hairless except for two patches of black hair at the temples, and had dark violet eyes with a wide iris. 5 feet 2 inches (1.5 metres) tall, he had white skin, without wrinkles, and looked to be in his forties or fifties. In fact, he told Diaz, he was eighty-five. The spaceman spoke at length with Dr Diaz, telling him that the extra-terrestrials are living among us to show us how to avoid the total destruction of our civilization. Misuse of energy sources, and pollution of the atmosphere, thus contaminating space as well as Earth, were reasons for their intervention. Diaz was instructed to proclaim the message the extra-terrestrial had given him:

> God is everywhere, we must live in God . . . all the religions you profess on earth, they are only roads to the same purpose – to know God. Instead of following so many roads, follow only one religion . . . the truth cannot be divided . . . the only way for you to defeat all the conflicts you have is to show love, which is the weapon that you have to defeat selfishness . . . there is nothing imperfect in the Cosmos. Everything is perfect, but in your mind you have the idea of imperfection . . . God is in our fellow human beings.[14]

If we believe what these temporarily Earth-dwelling extra-terrestrials say, there are incognito spacemen everywhere: UFO enthusiasts who immerse themselves in reports of contact and become convinced of the extra-terrestrials' great interest in Earth and what we are doing to it, are likely, if they are very impressionable, to begin seeing spacemen themselves. Some of them have even reported doing so, but such encounters are more likely to be the result of an overactive imagination, we suspect. It is understandable when a child, who has experienced strange events that sound like a UFO abduction, claims to have later seen space entities during her daily life. Próspera Muñoz, who was six or seven years old when abducted from her home near Jumilla in south-eastern Spain some time in the 1940s and medically examined, told how she saw the entities again afterwards. They were ugly and wore overalls or uniforms with turned-up collars. The later meetings took place around 1954 in her father's bar in Jumilla, in 1959 or 1960 on the beach at Alicante, in the Alicante telephone exchange in 1971, and in the Gerona telephone exchange in 1982, where she was working.[15] It is possible that the people she saw on these occasions and thought were

UFO entities simply reminded her visually of the beings she had encountered during her abduction, an event which had obviously fixed itself firmly in her memory.

The encounters with spacemen on Earth described here are only a few of the reports of such encounters which exist. A complete book has been written about the extra-terrestrials who are here on Earth to herald the New Age which is about to dawn – Ruth Montgomery's *Aliens Among Us*[16] – and who are we to deny that these events took place exactly as described? The difficulty with all these reports of human contact with extra-terrestrials is that they cannot be subject to scientific scrutiny and therefore cannot be proved either true or false. Only when the space people, if such there be, unequivocally reveal themselves to the naturally sceptical scientists, will their existence be undeniable. And so we must all continue to rely on our own personal intuition regarding the truth or otherwise of the data presented.

14 Is Man alone in the Universe?

Some concluding thoughts, by Janet Bord

I realize that my answer 'Yes' to the question posed in the chapter title must be accompanied by the acknowledgement that I am not possessed of a unique insight granted to me alone: in fact *no one* on Earth knows whether there is other intelligent life in the Universe. All we have are our beliefs, based (sometimes) on logical reasoning, and what follows is simply the way I personally see things.

Origins of life

I differentiate between 'life' and 'intelligent life': it is highly probable that 'life' exists elsewhere, but improbable that 'intelligent life' does. First, some evidence supporting the existence of life in outer space.

There is continual interaction between the planet Earth and outer space, in the form of meteorites (small pieces of rock which fall to Earth from space) and micrometeorites (which are tiny particles possibly from comets, and which are too small to be seen but of which around 10^7 kg per day in total settle slowly on the surface of the Earth). Amino acids have been found in meteorites: analysis of the Murchison meteorite that fell in Australia in 1972 showed the presence of sixteen amino acids, which are needed for the formation of proteins. Scientists were convinced that terrestrial contamination of the meteorite had not occurred. Research into other meteorites has resulted in equally tantalizing discoveries. Although much of the meteoritic material falling to Earth is burned up in the atmosphere at temperatures around 10,000°C, it seems clear that plenty survives, either in the form of solid rock or as dust, and that this material can contain the basic building blocks of life. Although we cannot see it, an organic rain is falling on us from space; and it is believed that 4,000 million years ago the bombardment could have been much more intense, with the result that Earth was 'seeded' with the basic requirements for life to form.

Comets, bright bodies orbiting the Sun and consisting of a frozen nucleus made up of water, methane and ammonia, with a tail of dust particles and gases, may also have an important role to play. Research is continually in progress into the make-up of comets and their behaviour, and in 1987 the first discovery was made of organic matter in a comet which was paying a visit to our Solar System. Comet Wilson was observed by astronomers through the Anglo-Australian Telescope in New South Wales, and the confirmation of its organic matter helps boost the theory that the carbon-based chemicals from which life on Earth evolved could have been brought here by comets.[1] Another recent theory is that comets may bring water to the Earth, indeed that the oceans were formed by water from comets, which is being continuously 'topped up'. One scientist has estimated that if only 10 per cent of the incoming mass consisted of icy comets during the vital time 3.8–4.5 billion years ago, the Earth could easily have acquired all its ocean water by cometary bombardment.[2] In support of this theory, it has been discovered that the water in Halley's Comet has the same abundance of two key isotopes as the Earth's oceans.

There is also considerable evidence being accumulated to support the theory that the cometary bombardment is still continuing. For example, physicist Clayne Yeates has located and photographed, using a telescope with a moving field of view, fast-moving objects 8–16 feet (2.4–4.8 metres) across near the Earth, which are believed to be previously undetected comets.[3] Yeates captured the images at the rate of about one every minute, showing that they enter the Earth's atmosphere by the millions every year, vaporizing there. His work followed that of physicist Louis A. Frank, working at the University of Iowa where from 1981 to 1986 he and his team studied thousands of observations from a research satellite in polar orbit 14,500 miles (23,335 kilometres) above the Earth. The team found about 30,000 small, mysterious black spots in ultra-violet images of the Earth. They concluded that these were clouds of water vapour from small icy comets that had broken up as they entered the atmosphere.[4]

So if, as seems likely, the essentials for the development of life arrived on Earth from outer space, is it not possible that the same essentials have also been deposited on other planets in our Solar System, and on planets elsewhere in our galaxy? It is even possible that the kinetic energy in the flux of small comets is sufficient to increase a planet's temperature as well as the comet supplying water and the essentials of life to that planet, life thereby developing on a planet which would originally have been too cold to support it. And not only comets may be responsible for carrying life to other planets. It has been suggested that the impact of a large meteorite on Earth, big enough to excavate a crater 60 miles (96 kilometres) wide, would dislodge small chunks of the Earth's crust which could then carry viable micro-organisms into interplanetary space. If the rocks were large enough, the micro-organisms would be shielded from ultra-violet radiation, low-energy cosmic rays, and even galactic cosmic rays, and could remain viable for a long time. S. A. Phinney and colleagues at the University of Arizona calculated in 1989 what would happen to 1,000 particles ejected from the Earth in random directions. 'Of the 1,000 hypothetical particles, 291 hit Venus and 165 returned to earth; 20 went to Mercury, 17 to Mars, 14 to Jupiter and 1 to Saturn. Another 492 left the Solar System completely, primarily due to gravitational close encounters with either Jupiter or Mercury that "slingshot" them on their way.'[5] There are several large meteor craters on Earth's surface, so it is entirely feasible that the Earth long ago exported life to outer space.

Knife-edge balance of life on Earth

However, the dumping of the seeds of life on a planet is one thing; the eventual emergence of intelligent life is quite another. I do not intend to trace the development of intelligent beings on Earth from the earliest stirrings of life; let us simply assume that it somehow happened, and here we are today. That 'somehow' includes many vital stages of development which it is likely the scientists do not yet understand, but there are certain criteria which they are aware of, and the listing of just a few of these should serve to illustrate how tenuous our existence really is.

First the stability of the Solar System is vital. If the Sun suddenly burned more fiercely, we would shrivel; if it suddenly became cooler, we would freeze. We are dependent on the Sun not simply for light, but for our very existence. So if the Earth were orbiting either closer to or further away from the Sun, terrestrial conditions would be hotter or colder and life as we know it would not exist. As has already been made clear in Chapter 2, the planets which do not have the Earth's advantages have been shown to be lacking in intelligent life; indeed there are few obvious signs of any life at all having ever existed on the other planets in the Solar System. Even life on Earth has to be protected from the Sun's harmful ultra-violet radiation, and, amazingly, that protection exists in the form of the ozone layer, at 9–18 miles (15–30 kilometres) high in the Earth's upper atmosphere. If the ultra-violet radiation were to reach the Earth's surface it would be biologically damaging: crops would give lower yields or a lower nutritive quality, plankton and other ocean life essential to the marine food chain would be damaged, and the end result would be global ecological disaster and famines. In recent years it has become clear that Man's activities are damaging this vital ozone layer, and we may soon learn at first hand the reality of its importance to life on this planet.

Life on Earth is also dependent on the existence of a satellite, the Moon. Scientists believe that intelligent life could not have developed here had not the Moon's gravitational pull given rise to oceans and continents and tectonic plates. The strong magnetic field which developed when the Earth's core melted due to tidal friction caused by the Moon, helped to protect emerging life from space radiation.[6] The Moon also helped stabilize the orientation of the Earth's rotation axis in space; without this stability long-term climatic changes would have resulted, killing off life on Earth. So it seems that the Earth's size, the Moon's size, their relationship and distance from the Sun, and the Sun's own stability, are all vital criteria which led to the formation of first life, then intelligent life, on Earth. It is unlikely that these events have all come together again elsewhere in the galaxy, maybe even the Universe – not impossible, I admit, but unlikely.

But, as if by a miracle, they have happened once, and intelligent life has formed. Not only are we reliant on the Sun, Earth and Moon remaining stable, we are also reliant on the mix of gases which make up our atmosphere. We do not often think about the air we breathe, because we cannot see it, but like a fish out of water we would soon know if our air supply was taken away. Not only we, but all the animals and plants, live in a close, even interdependent, relationship with each other and the planet, a relationship we call 'nature', and we tamper with the balance of nature at our own peril. Indeed the signs are now with us that we are upsetting the vital balance of nature. One result is that an excess production of carbon dioxide is believed to be leading to a 'greenhouse effect' whereby the temperature of the atmosphere will rise, causing in the first instance the melting of polar ice leading to a rise in sea levels and flooding, and also causing the failure of crops in already marginal areas as well as changes to the flora and fauna around the globe. In the long term, the temperature on Earth would eventually reach 900°F (480°C), like that on Venus, and life would be no more.

Such a danger as this is entirely man-made and could be avoided, yet we also face other threats to our existence which we cannot control. These could be caused by asteroid impacts, which fortunately are exceedingly rare but could happen at any time. Meteoritic debris on the floor of the south-eastern Pacific Ocean indicates that an object at least a third of a mile (half a kilometre) in diameter smashed into the Earth 2.3 million years ago, at a time when there seems to have been a major deterioration in the Earth's climate. The meteorite impact may have vaporized enough water into the atmosphere to cause the reflection of sunlight back into space, lowering the temperature and triggering the Ice Ages.[7] Another asteroid or comet impact about 65 million years ago is thought to have caused a global fire. The evidence was in the form of soot formed from the burning of vegetation, calculated as equivalent to 10 per cent of all today's plant material.[8] The extinction of the dinosaurs about 65 million years ago has also been blamed on a comet, the resulting change in climate depriving the animals of their food.

It has even been suggested that the development of advanced life on Earth was delayed because of asteroids continually striking the planet and wiping out the early life-forms. Professor James Kasting of Pennsylvania State University said, 'Asteroids that vaporized the oceans must have sterilized the entire planet, killing any organisms that had managed to evolve since the last such object struck.' Asteroid collisions seem to have been a more common event millions of years ago than they are now (fortunately for us), and Professor Kasting's resultant theory is that the delay between the appearance of single-celled organisms 3,500 million years ago, and the appearance of mammals and reptiles 500

million years ago, was caused by the asteroid collisions. A huge asteroid would vaporize the oceans, and raise the temperature of the atmosphere to 3,000°F (1,650°C), and the resulting steam would take at least a thousand years to dissipate and the oceans to re-form.[9]

Although these major events took place in the distant past, there is no reason why we should not ourselves experience a cometary impact at any time, with results which could be devastating, perhaps resulting in the extinction of intelligent life on Earth. Comet Encke and its accompanying cloud of asteroid debris is one visitor we need to be wary of. It passes through Earth's orbit once every 3.3 years, at which time some of the debris could collide with Earth. That may be what occurred in 1908, when an explosion blasted 2,500 square miles (4,000 square kilometres) of forest in Siberia. Astronomers calculate the culprit was a block of ice 40 yards (36 metres) wide and weighing about 30,000 tons. Apart from Encke, there are 10,000 million comets in the Solar System, as well as

Huge meteorites have collided with the Earth in the past, and could easily do so again. Craters like this one, in Arizona, demonstrate how vulnerable our planet is to potential disasters which could overwhelm Mankind.

assorted large asteroids orbiting close to Earth, like Hermes which came within 500,000 miles (800,000 kilometres) of Earth in 1937.[10] In March 1989 another asteroid came equally close to colliding with us. It weighed about 400 million tons, and if it had hit the Earth, 100 million people could have been killed. As Dr Henry Holt, the astronomer who discovered the asteroid, commented, 'If it had appeared only a few hours earlier it would have nailed us.' Astrophysicist Dr Victor Clube said, 'If it had crashed on the land it would have made a crater about 30 miles [48 kilometres] across. It would have thrown up huge clouds of dust into the atmosphere, blocking out the Sun's warmth.' Dr Bevan French, NASA planetary expert, added, 'If it had struck the sea its impact would have created waves nearly a mile high that would have devastated surrounding continents. This object crosses Earth's orbit so frequently that a collision is certain within the next five million years or so. But the danger lies in the existence of twenty or thirty other asteroids, whose orbits we do not know precisely.'[11]

We also risk being blasted by radiation from an exploding supernova, which occurs when a giant star flares up. We can expect such an event to take place within fifty light-years of Earth every few hundred million years or so, not a very great danger, but the consequences would be fatal for life on Earth.

Man is unique

With so many dangers facing us, both originating in outer space and with our own terrestrial activities, it is surprising that advanced life-forms have survived for as long as they have on this planet. Since other planets face similar dangers, and in addition do not have all the advantages for life that Earth has, it seems highly unlikely that a life-form similar to Man could have evolved elsewhere. I believe that Man is unique; but not that he is in any way special or more important than the other life-forms that have evolved on this planet as a result of the very precise conditions that obtain here. Indeed, it has been seriously suggested that it is merely due to luck that Man developed at all: 'Most people assume that evolution is progressive, that it leads inevitably to us, and that humans are the crowning achievement of life on Earth.' But, as the man who made that statement, Stephen Jay

Gould, points out in his book *Wonderful Life*, it is more likely that chance plays a greater role in evolution than is generally accepted. Marine fossils from the Burgess Shale, in a mountain quarry in British Columbia, include some of the best-preserved and oldest examples available for study, and palaeontologists have reconstructed from them some weird and wonderful creatures like nothing now living on Earth. They did not survive; others did; but the losers were not inferior to the survivors, and it could just as well have been Man's remote ancestors that died out.[12] Then intelligent life would not have developed on Earth – or would it have taken some other form? We can only speculate.

The lesson from this is that Man is just one of many life-forms on this planet. Because of the way he has developed, he (unfortunately) holds power over other life-forms, but this power does not make him superior. In countless ways Man is inferior to other creatures, even the humblest; we should remember that every species is special in its own way; each has its own niche on the planet. Just as sea-creatures are highly specialized and can only live in their particular watery environment, so is Man highly specialized and only equipped to live in the environment that gave birth to him. It follows therefore that I do not see any future in Man's plans to travel to other planets and build space colonies. Outer space is a hostile environment to Man, and he can only live there if he takes with him a self-contained earthly environment in which to live. The expense of doing this on any scale will bankrupt any Earth-dwelling civilization, and anyway what's the point? We already live on the most beautiful planet there is, and it seems to me we would be better employed spending the money on keeping it that way.

There is much that Man cannot know. Our knowledge is already vast, but even so we are only scratching the surface of total knowledge. We know much about outer space, but still there is much to learn about the tiniest life-forms on this planet. The more one probes into what may seem to be the minutiae of life on Earth, when compared with the larger questions relating to the Universe, the clearer it is that life is a highly complex affair, so much so that it has led many people to believe that all life was created by a superior being whom we call God. This concept certainly helps to subdue the uncertainties that thinking about the incredible diversity of nature will invariably induce: this just couldn't have happened at random – there must be a creator behind it. Not wishing to become embroiled in religious argument, I will not begin to discuss the arguments for and against this concept, and all the beliefs that have derived from it. I will only suggest that the idea of the creator as a human-like being, masculine of course, which has grown up over the centuries is misleading and should be discarded in favour of a more esoteric yet rational concept: that nature itself is God. Since the nature of which we are a part is unique to planet Earth (it has to be, given the unique circumstances, described earlier, in which life was able to develop on Earth), then it follows logically (*I* think) that Man too is unique to planet Earth.

Who are the UFO entities?

If the last statement is correct, where does that leave all the people who claim to have spoken with humanoid entities from planets X, Y and Z? Are they all telling lies, experiencing hallucinations, or is there some other explanation? I believe some are almost certainly liars, others are hallucinating or victims of their vivid imaginations. Some may be driven by a deep subconscious need to express the fear that lies within many people about the state of the planet, and what Man is doing to his environment. Many of the 'space people' who spoke to the contactees warned them of the error of our ways, naming in particular atomic weapons testing and other nuclear pollution. It is natural that as a species we should feel concern about what is happening, and it may be that in some people this concern expresses itself in a symbolic drama: a more advanced and benevolent being from outer space comes to Earth and speaks to humble Man, warning him about the dangers and telling him to spread the word among his fellows. In this way the contactee, consciously or unconsciously, believes himself to be doing something to make Mankind more aware of the danger he is in.

Although the tales told by the contactees may sometimes seem ludicrous, running through them there is often a thread of deadly seriousness which emanates from our collective unconscious, and which we ignore at our peril. In recent years the messages have become even more urgent. Instead of meeting friendly spacemen, humans are now being abducted by decidedly unfriendly space aliens and subjected to medical examinations against their will. Does this symbolize our fear that the planet is being raped, with ourselves as individuals feeling powerless to do anything about it? There is at present a dichotomy within the UFO world, with some researchers taking all such cases at their face value, and others interpreting them as symbolic events.

My feeling is that both the contactees' and abductees' stories are unlikely to depict real happenings; but there may be a third category of witnesses who genuinely did meet alien beings, especially among the people in Chapters 6–8 who claimed only one encounter with UFO entities. I think it highly unlikely that the alien beings these people have seen came from planets in our Solar System or from planets elsewhere in our galaxy or even other galaxies. Even if they had discovered a method of instantaneous travel from a point light-years away, many questions remain to be answered: Why come here and behave in so pointless a fashion over several decades? Why are so many aliens apparently coming from so many different sources to visit planet Earth? Why are they all basically humanoid in shape, when each planet with its different living conditions would express its life-force through living forms uniquely suited to those conditions? If we believe that alien beings would be travelling here as frequently as the UFO reports (not just those reports in which aliens are sighted) suggest, we are giving our tiny planet an importance it does not merit, in view of the vast number of other Solar Systems and planets that there are in the Universe. If we believe that alien beings nearly identical in appearance to ourselves are coming here, we are stretching probability to its furthest limits.

But if we consider that these visitors might be travelling much shorter distances to reach us, indeed no distance at all, then there might be more reason behind their appearances. Where would they originate from, if they did not need to travel through space to get here? The concept of a parallel universe has always appealed to me, as for one thing it is a tidy way of solving so many mysteries: all manner of entities could slip through the curtain dividing our world from theirs, mysterious communications could originate there, miscellaneous unexplained happenings reported by apparently sane people could be attributed to an unexpected glimpse through the curtain. I realize that in postulating the existence of a parallel universe (or even universes) whose inhabitants seem to be free to visit us at will, I have totally abandoned my attempt to solve the mystery of encounters with UFO entities by means of logic and common sense; but I do not believe the identification of UFO entities as extra-terrestrial beings to be logical either. The alternative to dismissing the reports totally is to venture into the murky world of fantastic theories, and so I put my toe into the water with trepidation.

We obviously cannot rely on what the entities themselves are telling us about their origins: they may have reasons of their own for concealing their true source from us. If it is as close as I suspect, they may fear that we could, if we worked on it, find our own way on a regular basis into their world, a prospect they would not welcome. If this theory were indeed correct, the implications would be quite serious, the main one being: is Mankind being secretly controlled by alien forces? By entering into involved discussion of this theory (and I stress it is only a theory), I would be moving too far away from the main subject of this conclusion, which is whether Man is alone in the Universe. To sum up my beliefs on the UFO entities: I believe that any who have a reality outside the imaginations of their human witnesses are *not* from outer space, even though they might say they are, and that their true home is much closer at hand.

So far as life on other planets is concerned, I believe it could, probably even does, exist. Simple life-forms are probably legion. Whether they might have anywhere developed into complex forms, even intelligent forms, is much more difficult to decide. We know so little about what is possible, despite our already encyclopaedic scientific knowledge, that it is risky to be too dogmatic, but my own feeling is that any complex life-forms that have developed elsewhere would not be intelligent in the way we determine intelligence. They would indeed be so different from us that we would probably not even be able to recognize them as 'intelligent'. Hence the likelihood of us being able to communicate with them in any way whatsoever would be negligible. And so that really is my answer to the question in the title of this chapter: Yes, we are alone in the Universe – when it comes to intelligent beings with whom we could communicate.

Some concluding thoughts, by Colin Bord

Having presented such evidence as we have been able to muster relating to the question, 'Is there life beyond planet Earth?', I must now attempt to formulate an answer, examining various aspects of the question and seeing where they lead me.

On the astronomical front there is great activity being directed to searching for signals from space. Evidently some influential scientists believe in a strong probability of life out there and are prepared to direct large sums of money towards searching for intelligent signals. There is, however, no reason to assume that advanced intelligences with appropriate technologies are likely to be using radio wavelengths, such as we use, with which to communicate. And if they do monitor such wavelengths in order to detect emerging primitive technologies, they may very well prefer not to reply, being merely content to continue monitoring and observing a burgeoning life-form. The odds against two such emerging technologies developing the use of the same radio wavelengths at the same time and thereby contacting one another are so vast as to make the occurrence virtually impossible. Even if communication were established by this means to anywhere other than the closest stars, the time-lag factor would make any sensible form of question and answer unusable. A question sent from Earth would receive an answer several generations later, and the recipients, whose scientific knowledge would presumably have advanced in the meantime, might well decide that the original question had not been well formulated. So although the search for extra-terrestrial intelligence through radio wavelengths is an interesting concept, I do not expect it to produce anything other than a negative result.

What of the hundreds, perhaps thousands, of people who believe that they have seen and sometimes spoken with spacemen emerging from landed craft and moving about on Earth? I don't doubt that a large proportion of these witnesses are honestly attempting to describe what they think they have experienced. But what exactly did they experience? At each of these events there is never a second witness to confirm the story, and where nearby individuals might be expected to have seen or heard something to suggest that an unusual occurrence was happening, they do in fact see and hear nothing untoward. It is an intensely personal inner-directed experience, but that is not to say that it is totally unreal or of a subjective origin. There is one school of thought which speculates that the entire phenomenon is a product of psychological traumas, but is this not simply seeking an easy blanket explanation? When we move on to consider the more complex cases presented by the contactees, we find the same forces at work, producing an intense experience which is primarily of significance to the particular individual involved, and leaving them with what is ultimately only anecdotal evidence. Have these witnesses been in contact with a being from Mars, Venus or some more distant planet? If they have, then I suggest it has not happened within our familiar world of three-dimensional space and time but in a further dimension into which the entities can draw the percipients.

Parallel universes

The idea of other dimensions existing, beyond our familiar three-dimensional universe, is no longer in the realms of science fiction but is a matter for discussion among today's physicists and mathematicians.[13] This other universe does not exist somewhere 'out there', but is right here beside us, its energy vibrating at frequencies we cannot detect, and yet if its denizens have mastered the art of moving in and out of our three dimensions, many of the strange phenomena we have encountered in earlier chapters might be explained. In fact, physicists say that there is insufficient mass in our Universe to explain the gravitational effects which can be measured and that to explain these effects 90 per cent of the matter which exists must be hidden from us, enough matter in fact to suggest the existence of not just one other universe but a multitude of parallel universes. But why should the beings of the parallel universes wish to contact humanity, and yet keep their true origins hidden?

Perhaps we should not lightly dismiss the frequent warnings which these entities give the percipients, regarding the repeated testing of nuclear devices which have 'upset the balance' of the planet or Solar System or even galaxy. Such subatomic reactions may be producing unwanted effects in parallel universes of which we have as yet no conception, and the concerned 'spacemen' may be travelling from dimensions which are much closer to us than even Mars or the Moon. There are repeated hints and suggestions that they are not of our physical level but of a more rarefied 'vibration', intangible to us, but totally solid and real themselves, and that they do not wish to interfere with the Earth's inhabitants but must sometimes interact with us for their own self-protection, and this includes misleading us about their true origin. They may reason that it is far better, with such an aggressive species as Mankind, to suggest that they are from an impossibly distant planet, rather than that their world is very close to our own, and one whose entrance could be revealed by further research into quantum mechanics. Better perhaps that Mankind remain unaware that there is a door leading to other levels of existence. It may also be conceivable that these other-dimensional worlds also encompass the whole of our Solar System, and that in a sense the 'space beings' do therefore also come from Mars, Venus or wherever, though not the Mars and Venus as seen through terrestrial telescopes. Living outside our space and time as they do, for the 'space people' the distances between all our Solar System planets may be negligible or even non-existent.

Although the nuclear threats and tests have lessened in recent years, there are still occasional reports which suggest that the entities continue to observe us and are still concerned with our behaviour. We have a report from as recently as 1988 about a man in rural Gloucestershire who held a short conversation with an entity one night in his back garden, via an unmanned probe hovering overhead. During the conversation two interesting points were made by the hovering voice, first: 'You are gaining knowledge so fast and getting to the very basis of the structure of matter and you could cause untold harm if you don't know what you are doing', and secondly, when asked where they were from, the voice was characteristically evasive, answering the suggested outer-space points of origin with 'We may be' and adding, 'There's no point in asking such questions', suggesting that these entities still feel a potential threat from humanity and will not yet reveal their real origins.[14] I expect that similar contacts may be on the increase during the 1990s, but will the messages have a different slant from those of the 1950s and 1960s?

Today in the more relaxed international climate of the 1990s humanity has other, but equally pressing, threats to its well-being in the many forms of pollution of air, land and sea which threaten us, in the contaminated food we eat, and the foul water we drink. Contactee reports virtually ceased by about the early 1970s, but what if they were to recommence today? Would the nuclear warnings be replaced with environmental ones? It is perhaps worth noting that not infrequently in the early contactee cases the virtues of a natural, wholefood, fresh and vegetarian diet were stressed, though in those years such ideas did not have the same significance to most people as they have today, and this aspect of the contacts did not receive any particular attention at the time. Another recurring theme which the contactees report is the concept of time. They are told, 'Time does not exist', or 'We do not exist within time', or 'We are not limited by time', or as Aura Rhanes said to Truman Bethurum, 'What you call time and distance is inconsequential in our lives.' It is worth pointing out that when these contact messages were received, quantum mechanics and its associated theories were quite undeveloped and in their infancy, and the contactees were people who were not likely to be familiar with such little-published abstract mathematical concepts.

Tenacity of life

To assume that life on planet Earth is an accident brought about by a series of coincidences that could never occur again is to view the Universe as the result of a chain of random reactions, confined within a framework of three-dimensional laws. A study of life on Earth shows that nature is prolific, the life force is abundant and will immediately colonize any available niche and start to proliferate under the most adverse conditions. The lowly bacteria have been found thriving just about everywhere. At great depths on the ocean bed far below the levels at which sunlight and heat can penetrate there are volcanic vents of hot water gushing from the Earth's interior. In these hot vents bacteria live using the heat and mineral content of the upwelling water to sustain life. Likewise some of the boreholes sunk into the Earth's surface, to a depth of 1,610 feet (520 metres), reveal that at all levels down to the greatest depths living bacteria are to be found, not as isolated oddities but in great numbers: more than 3,000 types of micro-organisms have been identified, many of them previously unknown to science.[15] They are also found in the active cores of nuclear reactors, and may present a future problem when radioactive waste has to be stored for thousands of years: over long periods they may dissolve casing materials and transport and precipitate radioactive material.[16]

Samples taken from the upper atmosphere also record an abundance of bacteria and it seems very possible that micro-organisms are continually arriving on Earth from space. Two professors of astronomy, Sir Fred Hoyle and Chandra Wickramasinghe, have plotted flu epidemics against sunspot activity using data going back to 1761 and have found a correlation for the last seventeen eleven-year sunspot cycles. (The sunspot cycle is the recurring eleven-year period during which the surface activity of the Sun rises to a peak of sunspot activity and solar flares, sending a 'solar wind' of charged particles into space. It is theorized that this solar wind carries the flu virus into the Earth's upper atmosphere from where it eventually descends on to the surface to start an epidemic.)[17]

Hoyle and Wickramasinghe also theorize that many forms of bacteria and virus arrive on Earth in this way, and are also brought to Earth by means of comets. Space is not in fact a dead and sterile emptiness; it teems with life which is waiting ready for a suitable environment in which to establish itself. It is known that life on Earth depends on a multitude of variables being exactly right before it can flourish, and that though this has happened once it is unlikely to occur again on the billions of possible planets throughout this galaxy or any other. If we assume that random chance alone has permitted life to become established on Earth, then the odds against it occurring again are perhaps too great. But there is an alternative, and that is to suggest that superior intelligences may have engineered the conditions upon planet Earth and established and watched over the original life-forms. This is not to suggest the existence of the omnipotent 'God' creator of the traditional religions, but to think more in terms of space beings with far more power and wisdom than humanity has. Referring back to the 1988 incident of the UFO witness and his conversation with the space probe, a relevant part of the exchange started when the voice from the probe

The Earth rising, as seen from the Moon. Photographed by Apollo 8 astronauts in 1968, this image epitomizes Earth's fragility as our planet floats alone in space.

said, 'If your species don't learn some sense we will have to take measures against you.'

'What do you mean? Are you going to exterminate us?'

'No – nothing like that. We'd have to take action to reduce your activities.'

'Could you do that?'

'Oh yes, easily. Wouldn't take much to upset your order of things.'

'What would you propose to do then?'

'You are very susceptible to bacteria and viruses. Something like that would cause disorganization.'

'You could introduce this into our world?'

'Yes, of course.'

'How will we know it's you?'

'You won't know. We don't propose to tell you.'[18]

Perhaps the Ancients' belief that plagues were visited upon humanity by the gods contains more truth than we have hitherto realized.

Many people have a great resistance to accepting the possibility that intelligent life could exist beyond the confines of this planet. Just as it took several generations to accept the Copernican concept that the Earth was not the centre of the Universe, but did in fact orbit the Sun, so the concept of many life-sustaining planets, among the billion billion of possible planetary systems orbiting other stars, is gradually being assimilated by present generations. The further development of this concept, that many forms of intelligence may exist on these planets and in other-dimensional states too, will undoubtedly gradually be accepted by future generations. Perhaps when this state of Cosmic Awareness is eventually achieved, the space intelligences may feel that humanity is then ready to receive open contact.

Select Bibliography

Ashpole, Edward, *The Search for Extra-terrestrial Intelligence*, Blandford Press, London, 1989.

Bowen, Charles (ed.), *The Humanoids: A Survey of World-Wide Reports of Landings of Unconventional Aerial Objects and their Alleged Occupants*, Neville Spearman, London, 1969; Henry Regnery Company, Chicago, 1969.

Corliss, William R. (compiler), *The Moon and the Planets* (A Catalog of Astronomical Anomalies), The Sourcebook Project, PO Box 107, Glen Arm, MD 21057, 1985.

DiPietro, Vincent, with Gregory Molenaar and Dr John Brandenburg, *Unusual Mars Surface Features*, Mars Research, PO Box 284, Glenn Dale, MD 20769, 4th ed. 1988.

Evans, Hilary (compiler and editor), with John Spencer, *UFOs 1947–1987: The 40-Year Search for an Explanation*, Fortean Tomes, London, 1987.

Goldsmith, Donald, and Tobias Owen, *The Search for Life in the Universe*, The Benjamin/Cummings Publishing Co., Menlo Park, CA, 1980.

Hoagland, Richard C., *The Monuments of Mars*, North Atlantic Books, Berkeley, CA, 1987.

Hoyle, Fred, and Chandra Wickramasinghe, *Lifecloud: The Origin of Life in the Universe*, J. M. Dent & Sons, London, 1978; Sphere Books, London, 1979.

—— *Cosmic Life-Force*, J. M. Dent & Sons, London, 1988.

Jastrow, Robert, *Journey to the Stars*, Bantam Press, London, 1990.

Kerrod, Robin, *The Journeys of Voyager*, Prion, Multimedia Books, London, 1990.

Leonard, George, *Somebody Else is on the Moon*, David McKay Company, New York, 1976; Pocket Books, New York, 1977.

Moore, Patrick, *Mission to the Planets*, Cassell, Poole, 1990.

Murray, Bruce C., *Journey Into Space: The First Thirty Years of Space Exploration*, W. W. Norton, London, 1990.

Pozos, Randolfo Rafael, *The Face on Mars*, Chicago Review Press, Chicago, 1986.

Ridpath, Ian, *Messages from the Stars*, Fontana Books, London, 1978.

Steckling, Fred, *We Discovered Alien Bases on the Moon*, G.A.F. International, 1981.

Story, Ronald D., *The Space-Gods Revealed: A Close Look at the Theories of Erich von Däniken*, New English Library, London, 1976.

—— *Guardians of the Universe?*, New English Library, London, 1980.

—— (ed.), *The Encyclopedia of UFOs*, Doubleday & Company, New York, 1980; New English Library, London, 1980.

Vallee, Jacques, *Passport to Magonia: From Folklore to Flying Saucers*, Henry Regnery Company, Chicago, 1969.

—— *Confrontations: A Scientist's Search for Alien Contact*, Souvenir Press, London, 1990.

Wilson, Don, *Our Mysterious Spaceship Moon*, Dell Publishing Company, New York, 1975.

—— *Secrets of our Spaceship Moon*, Dell Publishing Company, New York, 1979; Sphere Books, London, 1980.

Notes

See Bibliography for publication details of those books not fully described here.

Chapter 1. Is contact feasible – or even desirable?

1. Zdenek Kopal, *Man and his Universe* (Hart-Davis, London, 1972).

Chapter 2. Life on Solar System planets

1. 'Wonderful News about Mercury', *Scientific American*, 62:211, 1890, quoted in Corliss, *The Moon and the Planets*, p. 39.
2. A. E. Douglass, 'The Markings on Venus', *Royal Astronomical Society, Monthly Notices*, 58:382, 1898, quoted in Corliss, *The Moon and the Planets*, pp. 339–40.
3. 'Another Flashing Lunar Mountain?', *Strolling Astronomer*, 10:20, 1956, quoted in Corliss, *The Moon and the Planets*, p. 130.
4. More information given in Corliss, *The Moon and the Planets*, pp. 210–12.
5. I. R. Wright, et al., 'Organic Materials in a Martian Meteorite', *Nature*, 340:220, 1989.
6. 'Is there life on Mars after all?', *New Scientist*, 31 July 1986, p. 19, noted in *Science Frontiers*, no. 48, p. 1.
7. Corliss, *The Moon and the Planets*, pp. 220–2.

8. ibid., p. 200.
9. *Applied Optics*, 27:1926, 1988, noted in *Science Frontiers*, no. 60, p. 2.
10. More information on the Martian structures can be found in DiPietro, Molenaar and Brandenburg, *Unusual Mars Surface Features*, Hoagland, *The Monuments of Mars*, and Pozos, *The Face on Mars*.
11. Adrian Berry, 'Titan's surprising far-off world', *Daily Telegraph*, 29 January 1990.

Chapter 3. Man sends messages into space – and receives some

1. Fred Hoyle and Chandra Wickramasinghe, 'Interstellar Bacteria', *Space Travellers: The Bringers of Life* (Cardiff, 1981), p. 75.
2. More information on organic compounds in meteorites can be found in William R. Corliss (compiler), *The Sun and Solar System Debris* (Sourcebook Project, Glen Arm, MD, 1986), pp. 201–4.
3. Colin Hughes, 'America pays to scan the airwaves for a message from ET', *The Independent*, 12 February 1990.
4. Jonathan Eberhart, 'Listening for ET', *Science News*, 135:296, 1989, noted in *Science Frontiers*, no. 64, p. 4.
5. A detailed account of the Tesla signals can

be found in Fred Bobb, 'Forgotten Tesla Letter Rediscovered', *Pursuit*, vol. 21 no. 1, whole no. 81, 1988, pp. 27–9.

6. Associated Press report from London, 23 August 1924.

7. Report from Vancouver, British Columbia, 21 August 1924.

8. *New York Times* reports from 23 and 28 August 1924, reproduced in *Pursuit*, vol. 21 no. 1, whole no. 81, 1988, p. 29.

9. Duncan Lunan's Boötes hypothesis is fully explained in his book *Man and the Stars* (Souvenir Press, London, 1974), issued as *Interstellar Contact* (Henry Regnery Company, Chicago, 1975) in the USA, and paperbacked in the States as *The Mysterious Signals from Outer Space* (Bantam, New York, 1977). Following the reaction of the scientific community to his book, Lunan wrote an updating chapter which was published in Chris Boyce's book *Extra-terrestrial Encounter* (David & Charles, Newton Abbot, 1979; New English Library paperback, London, 1981). Some criticism of Lunan's work can be found in Robert Scheaffer, *The UFO Verdict* (Prometheus Books, Buffalo, NY, 1981), pp. 129–32.

10. Associated Press bulletin published in US newspapers, 9 August 1988.

11. p. 378.

12. Report by Adrian Berry in *Daily Telegraph*, 17 November 1988.

13. Report by Adrian Berry in *Daily Telegraph*, 21 November 1988, describing a new book on radio astronomy by Professor Sir Francis Graham Smith and Professor Sir Bernard Lovell, *Pathways to the Universe* (Cambridge University Press, 1988).

14. Roger Highfield, 'Glimpses at the beginning of time', *Daily Telegraph*, 19 February 1990.

15. Report by Adrian Berry in *Daily Telegraph*, 18 January 1989.

16. Report by Marjorie Mandel in *St Louis Post-Dispatch*, 12 July 1988.

17. Report by Adrian Berry in *Daily Telegraph*, 11 October 1989.

18. Reports by Roger Highfield and Adrian Berry in *Daily Telegraph*, 22 December 1988; reports by Adrian Berry in *Daily Telegraph*, 21 July 1989 and 23 November 1989; Arthur E. Smith, *Mars: The Next Step* (Adam Pilger, 1989).

19. Report by Adrian Berry in *Daily Telegraph*, 29 November 1988, relating to the launch of Smith and Lovell's book *Pathways to the Universe*.

20. Report by Adrian Berry in *Daily Telegraph*, 11 October 1989.

21. Helen Gavaghan, 'Europe turns its sights towards Saturn', *New Scientist*, 26 November 1988, p. 18.

22. Adrian Berry, 'Cosmic wormholes: key to the universe', *Daily Telegraph*, 5 December 1989; John Gribbin, 'Time machines, wormholes and the Casimir effect', *New Scientist*, 15 October 1988.

Chapter 4. Have extra-terrestrials already visited Earth?

1. Souvenir Press 1970 edition, p. 118.

2. For more details and up-to-date information on research into the Nazca lines, see Nigel Pennick and Paul Devereux, *Lines on the Landscape*: Leys and Other Linear Enigmas (Robert Hale, London, 1989).

3. *Chariots of the Gods?*, Souvenir Press 1969 edition, p. 122.

4. Published 1974 in Germany as *Da tat sich der Himmel auf*; UK paperback published 1974 by Corgi Books. Edward Ashpole also discusses Ezekiel's visions in *The Search for Extra-terrestrial Intelligence*, pp. 147–9.

5. Sidgwick & Jackson, London, 1976.

6. More details can be found in Ian Ridpath, *Messages from the Stars*, Chapter 12.

7. *American Journal of Science*, 1:2:144–6,

1820, quoted in William R. Corliss
(compiler), *Strange Artefacts*
(Sourcebook Project, Glen Arm, MD,
1976), item MET-003.

8. J. B. Browne, *American Journal of
Science*, 1:19:361, 1831, quoted in
William R. Corliss (compiler), *Ancient
Man*: A Handbook of Puzzling Artefacts
(The Sourcebook Project, Glen Arm,
MD, 1978), p. 657.

9. David Brewster, Report of the British
Association, pt. 2, 51, 1844, quoted in
Corliss, *Ancient Man* op. cit., pp. 651–2.

10. *The Times* (London), 22 June 1844,
noted in Corliss, *Strange Artefacts*,
op. cit., item MET-001.

11. Report in *Scientific American*, noted by
Chris Cooper in 'Inadmissible Evidence',
The Unexplained, issue 40, p. 795.

12. J. Q. Adams, *American Antiquarian*,
5:331–2, 1883, quoted in Corliss, *Ancient
Man*, op. cit., pp. 653–4.

13. Chris Cooper, op. cit., p. 795.

14. Harry V. Wiant, Jr, *Creation Research
Society Quarterly*, 13:74, 1976, quoted in
Corliss, *Ancient Man*, op. cit., p. 654.

15. pp. 96–102.

Chapter 5. Are extra-terrestrials visiting us now?

1. Kenneth Arnold and Ray Palmer, *The
Coming of the Saucers* (Amherst Press,
WI, 1952), pp. 9–13.

2. *National Enquirer*, 1 June 1976.

3. Pamela J. Macleod, 'World Beats Path to
Martian Shrine', *Wall Street Journal*,
28 October 1988.

4. A range of these activities is described and
illustrated in Douglas Curran's book *In
Advance of the Landing: Folk Concepts of
Outer Space* (Abbeville Press, New York,
1985).

5. Associated Press report, February 1988.

6. Report by Bruce Maccabee, who
performed a detailed investigation, in
Story, *Encyclopedia of UFOs*, pp. 223–6;

computer analysis undertaken by William
Spaulding, see his article 'Analysing the
Trent Photos' in *The Unexplained*, issue
34, pp. 674–7.

7. Story, *Encyclopedia of UFOs*, pp. 366–9.

8. Steuart Campbell, 'UFO: Hoax or
Mirage?', *British Journal of Photography*,
15 June 1989, pp. 17–19.

9. Chris Rutkowski, 'Burned by a UFO?
The story of a bungled investigation',
International UFO Reporter, vol. 12
no. 6, 1987, pp. 21–4; letter from Edward
M. Barker, investigator in the Michalak
case, in *International UFO Reporter*,
vol. 13 no. 2, 1988, pp. 21–2.

10. J-J. Velasco, 'Scientific Approach and
Results of Studies Into Unidentified
Aerospace Phenomena in France',
MUFON 1987 International UFO
Symposium Proceedings, pp. 56–7.

11. Todd Lohvinenko, 'A Mysterious
Object', Royal Astronomical Society of
Canada, National Newsletter, 77:L19,
1983, quoted in *Science Frontiers*, no. 32,
p. 2. Many other cases are given in
William R. Corliss (compiler), *The Sun
and Solar System Debris* (The
Sourcebook Project, Glen Arm, MD,
1986), pp. 107–15.

12. Examples given in Corliss, *Sun and Solar
System Debris*, pp. 115–17.

13. Frank C. Clark, *Strolling Astronomer*,
10:67–8, 1956, quoted in William R.
Corliss (compiler), *Mysterious Universe:
A Handbook of Astronomical Anomalies*
(The Sourcebook Project, Glen Arm,
MD, 1979), pp. 524–5.

14. R. A. F. Edwards, 'Unidentified Flying
Object', *Marine Observer*, 54:82, 1984,
quoted in *Science Frontiers*, no. 38, p. 1.

15. Walt Andrus, 'Space Shuttle Discovery
Voice Recording', *MUFON UFO
Journal*, no. 255, July 1989, pp. 11–13.

16. List published by The UFO Contact
Center in their publication *The Missing
Link*, March 1989.

17. Mark Rodeghier, 'Roswell, 1989', in

International UFO Reporter, vol. 14
no. 5, 1989, pp. 4–8.

18. Ron Schaffner, 'Roswell: A Federal
Case?', *UFO Brigantia*, summer 1989,
pp. 12–16.

19. Kim Opatka, 'Kecksburg Crash
Controversial', *The Latrobe Bulletin*, PA,
6 May 1989.

20. Stan Gordon, 'The Kecksburg UFO
Crash', *MUFON UFO Journal*, no. 257,
September 1989, pp. 3–6, and
'Kecksburg Crash Update', *MUFON
UFO Journal*, no. 258, October 1989,
pp. 3–5.

Chapter 6. UFO entities from our Solar System

1. Coral E. Lorenzen, *The Great Flying
Saucer Hoax* (1962), paperbacked as
*Flying Saucers: The Startling Evidence of
the Invasion from Outer Space* (Signet
Books, New York, 1966), p. 57 of
paperback edition.

2. Gordon Creighton, 'Three More
Brazilian Cases', *Flying Saucer Review*,
vol. 13 no. 3, p. 5.

3. Aimé Michel, *Flying Saucers and the
Straight-Line Mystery* (S. G. Phillips,
Inc., New York, 1958), p. 154.

4. Dr Walter Buhler, 'Conversation with
entities at Marimbonda', *Flying Saucer
Review*, vol. 25 no. 3, pp. 18–19.

5. Coral E. Lorenzen, *Flying Saucers*,
op. cit., p. 53 paperback edition.

6. Both cases from Michel, op. cit.,
pp. 271–4.

7. Nigel Rimes, 'Another Hospital Visited',
Flying Saucer Review, vol. 15 no. 1,
pp. 4–6.

8. Daniel Cohen, *The Great Airship Mystery*
(Dodd, Mead & Company, New York,
1981), pp. 59–60.

9. 'The Spacemen threw stones', *Flying
Saucer Review*, vol. 7 no. 6, pp. 30–1,
reprinted from *APRO Bulletin*, May 1961.

10. 'Birmingham Woman Meets Spaceman',
Flying Saucer Review, vol. 4 no. 2,
pp. 5–6, and see also *Flying Saucer
Review*, vol. 5 no. 5, p. 5.

11. Mary King, 'Physical Contact in North
Devon', *Cosmic Voice*, no. 15, April/May
1958, pp. 5–8.

12. Berthold Eric Schwarz, M.D., 'Gary
Wilcox and the Ufonauts', *Flying Saucer
Review*, Special Issue 3, 1969, pp. 20–7.

13. Jenny Randles and Peter Warrington,
UFOs: A British Viewpoint (Robert Hale,
London, 1979), pp. 155–6.

14. Brad Steiger and Joan Whritenour, *Flying
Saucers are Hostile* (Tandem Books,
London, and Award Books, New York,
1967), pp. 119–30. Charles Bowen, 'Who
Hoaxes Who?', *Flying Saucer Review*,
vol. 11 no. 4, pp. 6–7.

15. Note 8 to Schwarz's article in *Pursuit*,
no. 82, vol. 21 no. 2, 1988, p. 51.

16. Berthold Eric Schwarz, M.D., *UFO
Dynamics*, Book II (Rainbow Books,
Moore Haven, FL, 1983), p. 351.

17. 'The Silpho Moor Mystery', *Flying
Saucer Review*, vol. 4 no. 2, p. 4; 'Silpho
Moor Controversy', *Flying Saucer
Review*, vol. 4 no. 4, p. 19, which has
photographs of disc and messages; Philip
Longbottom, 'The Silpho Moor
Mystery', *Flying Saucer Review*, vol. 4
no. 6, pp. 15–17, includes the complete
message; Jenny Randles, 'Mystery of the
Silpho Saucer', *UFO Brigantia*, no. 35,
November–December 1988, pp. 15–19.

18. Case 503 in Intcat, *MUFON*, New
Series 5, p. 8.

19. Desmond Leslie and George Adamski,
Flying Saucers Have Landed (Neville
Spearman, London, 1970, originally
published 1953).

20. (Neville Spearman, London, 1963),
p. 185.

21. Diane E. Wirth, 'Adamski on Trial',
Pursuit, no. 51, vol. 13 no. 3, p. 103.

22. Eileen Buckle, *The Scoriton Mystery*
(Neville Spearman, London, 1967).

23. Charles Bowen (ed.), *The Humanoids*, pp. 111–13.

24. ibid., pp. 118–19.

25. Leonte N. Objio and Richard W. Heiden, 'Humanoids in the Dominican Republic', *APRO Bulletin*, vol. 30 no. 1, pp. 5–6.

Chapter 7. UFO entities from Orion, Gemini, Zircon and other named places

1. Story, *Encyclopedia of UFOs*, pp. 402–4 has an article on 'Zeta Reticuli connection'.

2. John G. Fuller, *The Interrupted Journey* (The Dial Press, New York, 1966; Souvenir Press, London, 1980) recounts the Hill abduction case and subsequent investigation. Budd Hopkins has written two books on his own intensive research into abductions: *Missing Time* (Richard Marek Publishers, New York, 1981) and *Intruders* (Random House, New York, 1987). Philip J. Klass's book is a sceptical look at the subject: *UFO Abductions: A Dangerous Game* (Prometheus Books, New York, 1989). Jenny Randles, *Abduction* (Robert Hale, London, 1988) is a very readable and sensible survey. John L. Spencer, *Perspectives* (Macdonald, London, 1990) contains a critical review of the Hill case.

3. Given in full in Bowen, *The Humanoids*, pp. 200–38.

4. Luiz do Rosario Real, 'The New "A.V.B." Case from Brazil: Full Account', *Flying Saucer Review*, vol. 31 no. 3, pp. 16–24.

5. A. J. Gevaerd, 'The Abduction at Jardim Alvorada: Investigation under Hypnotic Time-Regression', *Flying Saucer Review*, vol. 30 no. 2, pp. 17–25.

6. *Saucer News*, vol. 16 no. 4, whole no. 74, pp. 9–10.

7. Gordon Creighton, 'A Weird Tale from the Vienna Woods', *Flying Saucer Review*, vol. 24 no. 6, pp. 20–1.

8. Gordon W. Creighton, 'The Italian Scene – Part 4', *Flying Saucer Review*, vol. 9 no. 4, pp. 10–11.

9. Richard W. Heiden, 'A 1949 Brazilian Contactee', *Flying Saucer Review*, vol. 27 no. 5, p. 28, and *Flying Saucer Review*, vol. 27 no. 6, pp. 19–20.

10. Gordon Creighton, 'The Rubem Hellwig Contact Claim', *Flying Saucer Review*, vol. 12 no. 6, pp. 27–9.

11. Horacio Gonzalez Ganteaume, 'Landing at San Pedro de los Altos', *Flying Saucer Review*, vol. 15 no. 2, pp. 21–3.

12. *Sandwell Express & Star*, 15 August 1989.

13. J. Antonio Huneeus, 'A Historical Survey of UFO Cases in Chile', MUFON UFO Symposium Proceedings, 1987, pp. 196–7.

14. Raymond E. Fowler, *Casebook of a UFO Investigator* (Prentice-Hall, NJ, 1981), pp. 149–53.

15. Irene Granchi, 'An Encounter with "Rat-Faces" in Brazil', *Flying Saucer Review*, vol. 29 no. 1, pp. 6–13.

16. Diana Dillaber, 'Alien Visitors?', *Oakland Press*, Pontiac, MI, 22 August 1982, reprinted in *UFO Newsclipping Service*, no. 159, October 1982, pp. 7–8.

17. Paul Devereux, *Earth Lights Revelation* (Blandford Press, London, 1989), pp. 112–13.

18. Gordon Creighton, 'A New Brazilian "A.V.B."', *Flying Saucer Review*, vol. 29 no. 4, pp. 9–11.

19. *UFO Report*, December 1979, p. 14.

20. Isabel Davis and Ted Bloecher, *Close Encounter at Kelly and Others of 1955* (Center for UFO Studies, 1978), pp. 184–6.

21. Jerome Clark, 'The Greatest Flap Yet? – Part IV', *Flying Saucer Review*, vol. 12 no. 6, p. 10.

22. Bowen, *The Humanoids*, p. 100.

Chapter 8. UFO entities from far-distant nameless worlds

1. Juan Jose Benitez, 'The Ufonaut's Plea for Water', *Flying Saucer Review*, vol. 24 no. 2, pp. 3–6; report by Cynthia Hind in *Afrinews*, no. 2, June 1989, pp. 24–31.
2. Richard Greenwell, 'The Experience of Señor C.A.V.', *Flying Saucer Review*, vol. 16 no. 6, pp. 12–13.
3. Full text of conversation between C.A.V. and Richard Greenwell published in Chapter 8 of Jim and Coral Lorenzen, *UFOs Over the Americas* (Signet Books, New York, 1968).
4. Gordon Creighton, 'A New South American "Wave"', *Flying Saucer Review*, vol. 14 no. 5, p. 23.
5. Jerome Clark, 'Two New Contact Claims', *Flying Saucer Review*, vol. 11 no. 3, p. 21.
6. Lorenzen, op. cit., p. 80.
7. Coral and Jim Lorenzen, *Encounters with UFO Occupants* (Berkley Publishing Corporation, New York, 1976), pp. 225–6.
8. Ralph Blum with Judy Blum, 'The Ordeal of Herbert Schirmer', in D. Scott Rogo, *UFO Abductions* (Signet Books, New York, 1980), pp. 112–21.
9. Timo Pyhälä, 'Contact in Helsinki', *Flying Saucer Review Case Histories*, no. 8, pp. 7–8.
10. K. Gösta Rehn, *UFOs Here and Now* (Abelard-Schuman, London, 1974), pp. 135–6.
11. Warren Smith, *UFO Trek* (Sphere Books, London, 1977), pp. 86–100.
12. Lawrence J. Fenwick, Harry Tokarz and Joseph Muskat, 'Canadian Rock Band Abducted?', *Flying Saucer Review*, vol. 29 no. 3, pp. 2–9.
13. Carl van Vlierden, 'Escorted by UFOs from Umvuma to Beit Bridge' and 'Investigation under Hypnosis: Contact Revealed', *Flying Saucer Review*, vol. 21 no. 2, pp. 3–10.
14. John H. Womack, *I was Picked up by a UFO* (The Helms Publications, Cullman, AL, 1975).
15. Gordon Creighton, 'A New Medicare?', *Flying Saucer Review*, vol. 24 no. 2, p. 9.
16. Nikita A. Schnee, 'Contact Reported near Pyrogovskoe Lake', *Flying Saucer Review*, vol. 26 no. 6, pp. 6–8, and Heikki Virtanen, 'Soviet Contact Case near Pyrogovskoe Lake – The Missing Pages', *Flying Saucer Review*, vol. 28 no. 3, p. 20.
17. Irene Granchi, 'Abduction at Mirassol', *Flying Saucer Review*, vol. 30 no. 1, pp. 14–22.
18. Fred Whiting, 'The Abduction of Harry Joe Turner', *MUFON UFO Journal*, no. 145, March 1980, pp. 3–7.
19. Lew Willis, 'Mother and Child Texas Abduction Case', *MUFON UFO Journal*, no. 167, January 1982, pp. 3–7.
20. 'A New Book on the UFO Situation in China', *Flying Saucer Review*, vol. 28 no. 6, pp. 22–3.
21. Janet Gregory, 'Similarities in UFO and Demon Lore . . . or, take off your shoes and show me your feet!', *Flying Saucer Review*, vol. 17 no. 2, p. 32.
22. Denise Lacanal and Théodore Revel, 'The Demon Who Admitted It! A Close Encounter in France in 1987', *Flying Saucer Review*, vol. 33 no. 4, pp. 6, 20–3.

Chapter 9. Extra-terrestrial contact by radio

1. George Hunt Williamson, *The Saucers Speak* (Neville Spearman, London, 1963), pp. 23–30.
2. ibid., p. 107.
3. ibid., pp. 106–7.
4. ibid., pp. 50–1.
5. ibid., p. 129.
6. ibid., pp. 73–4.
7. ibid. The story of Lyman H. Streeter is given in Chapter 11, 'Mr. "R.'s" Secret'.
8. Entry on John Otto in Margaret Sachs,

The UFO Encyclopedia (G. P. Putnam's Sons, New York, 1980), p. 234.

9. Bernard Smith, 'People from outer space contact earth man by "radio"', *Sheffield Star* report, reprinted in *Flying Saucer Review*, vol. 4 no. 4, pp. 28–9.

10. John Keel, *Our Haunted Planet* (Fawcett Publications, Greenwich, CT, 1971), p. 183.

11. The full text of this message can be read in Cynthia Hind's book *UFOs – African Encounters* (Gemini, Zimbabwe, 1982), pp. 93–103.

12. A series of articles by Bob Renaud on his contacts appeared in *Flying Saucers International*, beginning with issue no. 18 and ending with issue no. 29, March 1969.

13. Brad Steiger, *Mysteries of Time and Space* (Prentice-Hall, 1974; Sphere Books, London, 1977), pp. 139–40 of Sphere edition.

14. Andrija Puharich, *Uri* (Futura Publications, London, 1974).

15. Gordon Creighton, 'The Italian Scene once more', *Flying Saucer Review*, vol. 20 no. 2, p. 28.

16. This version of the message appeared in *Viewpoint Aquarius*, no. 66, January 1978.

17. Letter to the authors dated 14 August 1978.

18. *APRO Bulletin*, vol. 28 no. 8, p. 5.

Chapter 10. Contactees meet people from our Solar System

1. Howard Menger's experiences are described fully in his book *From Outer Space to You* (Saucerian Books, 1959; Pyramid Books, New York, 1967).

2. Entry on John Langdon Watts in Margaret Sachs, *The UFO Encyclopedia* (G. P. Putnam's Sons, New York, 1980), p. 363.

3. 'Luciano Galli's Contact Claim', *Flying Saucer Review*, vol. 8 no. 5, pp. 29–30.

4. Håkan Blomqvist, 'Some further notes from the Helge file', *AFU Newsletter*, no. 27, pp. 9–10.

5. Buck Nelson, 'A Strange Tale from Missouri', *Flying Saucer Review*, vol. 1 no. 1, p. 4; Warren Smith, *UFO Trek* (Sphere Books, London, 1977), p. 50.

6. 'Pie-maker Flew to Red Planet', *Flying Saucer Review*, vol. 4 no. 6, p. 4.

7. A. H. Matthews, 'I Knew Then That They Were Space Beings', the opening section of Matthews' narrative *The Wall of Light: Nikola Tesla and the Venusian Space Ship the X-12*, reprinted in John Robert Colombo, *Extraordinary Experiences* (Hounslow Press, Ontario, 1989), pp. 236–41.

8. Orfeo Angelucci, *The Secret of the Saucers* (Amherst Press, WI, 1955).

9. ibid., p. 42.

10. *Flying Saucers Have Landed*, p. 250.

11. The Adamski story is told in Desmond Leslie and George Adamski, *Flying Saucers Have Landed* (Neville Spearman, London, 1970 – first published 1953); George Adamski, *Inside the Space Ships* (Arco Publishers and Neville Spearman, London, 1956); George Adamski, *Flying Saucers Farewell* (first published 1961, published in paperback 1967 as *Behind the Flying Saucer Mystery*, Paperback Library, New York).

12. Lester J. Rosas, 'The Lester Rosas Story', *Flying Saucers International*, no. 29, March 1989, pp. 8–15.

13. Dan Martin, *The Watcher: Seven Hours Aboard a Space Ship* (Saucerian Publications, Clarksburg, WV, n.d.).

14. *Edge of Tomorrow, The Reinhold O. Schmidt Story* (privately published, 1963).

Chapter 11. Contactees meet people from Meton, Bâavi, Clarion, Zomdic and other far-away worlds

1. Elizabeth Klarer, *Beyond the Light Barrier* (Howard Timmins Publishers, South Africa, 1980).

2. Information on Monsieur Y's contact with Bâavi taken from the book *The Mysteries of the Skies*, pp. 277–86, author and publisher unknown.

3. Truman Bethurum, *Aboard a Flying Saucer* (DeVorss & Co., Los Angeles, 1954).

4. Warren Smith, *UFO Trek* (Sphere Books, London, 1977), p. 49.

5. 'Recent Contacts and Landing Reports', *Flying Saucer Review*, vol. 4 no. 4, pp. 26–7; also *Flying Saucer Review*, vol. 6 no. 2, p. 16.

6. Arthur Shuttlewood, *The Warminster Mystery* (Neville Spearman, London, 1967) and *Warnings from Flying Friends* (Portway Press, Wiltshire, 1968).

7. Wendelle C. Stevens and William J. Herrmann, *UFO . . . Contact from Reticulum* (privately published by Wendelle C. Stevens, n.d.); Wayne Laporte, 'The Charleston Close Encounters', *UFO Report*, December 1979, pp. 20–3.

8. Related by Woodrow Derenberger to author Harold W. Hubbard, *Visitors from Lanulos* (Vantage Press, New York, 1971); Karl T. Pflock, 'Anatomy of a UFO Hoax', *Fate* magazine, November 1980, issue no. 368, pp. 40–8.

9. *UFO . . . Contact from the Pleiades*, vol. 1 (Genesis III Productions, Phoenix, AZ, 1979); Gary Kinder, *Light Years: An Investigation into the Extra-terrestrial Experiences of Eduard Meier* (The Atlantic Monthly Press, New York, 1987); Wendelle C. Stevens, 'The Meier Contact: An Encounter of the Closest Kind', *UFO Report*, May 1978, pp. 20–3; Open letter from L. J. Lorenzen to Genesis III Productions, *The APRO Bulletin*, vol. 28 no. 2, p. 1; Kal K. Korff, 'The Meier Incident: The Most Infamous Hoax in Ufology' (William L. Moore Publications & Research, 1981); *Focus Newsletter*, Fair-Witness Project, Burbank, California, vol. II no. 12, 1987.

10. Dr Daniel Fry, *The White Sands Incident* (Best Books, Louisville, KY, 1966); Daniel W. Fry, *Alan's Message: To Men of Earth* (New Age Publishing Co., 1954).

11. John Keel, *The Mothman Prophecies* (Saturday Review Press, E. P. Dutton, New York, 1975), pp. 185–8; Warren Smith, *UFO Trek* (Sphere Books, London, 1977), pp. 17–20.

12. Claude Vorilhon, *Space Aliens Took Me to Their Planet* (Fondation pour l'Accueil des Elohim (FACE), Lichtenstein, 1978).

13. Margaret Sachs, *The UFO Encyclopedia* (G. P. Putnam's Sons, New York, 1980), p. 352; Bryant and Helen Reeve, *Flying Saucer Pilgrimage* (Amherst Press, WI, 1957), pp. 90–100; Gray Barker, *Gray Barker at Giant Rock* (Saucerian Publications, Clarksburg, WV, 1976), pp. 13–14; Long John Nebel, 'Contactees I Have Known', *The New Report on Flying Saucers* (Fawcett Publications), no. 2, 1967, p. 70.

Chapter 12. Space contact through mediumship and channelling

1. Emanuel Swedenborg, *Earths in the Universe* (Swedenborg Society, 1962).

2. Allan Kardec, *The Spirits' Book* (LAKE – Livraria Allan Kardec Editora, S. Paulo, Brazil), p. 77.

3. Warren Smith, *UFO Trek* (Sphere Books, London, 1977), p. 52.

4. Theodore Flournoy, *From India to the Planet Mars* (University Books, New York, 1963).

5. Harry Price, *Confessions of a Ghost-Hunter* (Putnam, London, 1936), ch. 8.

6. William Ferguson, *My Trip to Mars* (Saucerian Publications, Clarksburg, WV, n.d.).

7. Jerome Clark and Loren Coleman, *The*

Unidentified (Warner Paperback Library, New York, 1975), p. 213.

8. Allen Noonan, 'I went to Venus and Beyond . . .', *The New Report on Flying Saucers*, no. 2 (Fawcett Publications), p. 50; Rolf Alexander, M.D., *The Power of the Mind* (Werner Laurie, London, 1956).

9. Reverend George King, *You are Responsible!* (The Aetherius Press, 1961); George King, D.D., *Life on the Planets* (The Aetherius Press, 1966).

10. Margaret Sachs, *The UFO Encyclopedia* (G. P. Putnam's Sons, New York, 1980), p. 222; Ruth Norman, *Preview for the Spacefleet Landing on Earth in 2001 A.D.* (Unarius Academy of Science, El Cajón, CA, 1987).

11. Beth Kendall (compiler and editor), *The Evergreens: Visitors of Time and Space* (published by Michael Blake Read, Philippa M. Lee and Associates, Toronto, 1978).

12. Nigel Rimes, 'Muzio's Contacts', *Flying Saucer Review*, vol. 17 no. 1, pp. 24–6.

13. Ruth Montgomery, *Aliens Among Us* (Fawcett Crest/Ballantine Books, New York, 1985), pp. 98–127.

14. ibid., p. 48.

15. ibid., pp. 43–4; *Timothy Green Beckley's Book of Space Brothers* (Saucerian Publications, Clarksburg, WV, 1969), p. 20.

16. Stuart Holroyd, *Prelude to the Landing on Planet Earth* (W. H. Allen, London, 1977; paperback edition titled *Briefing for Landing on Planet Earth*, Corgi Books, London, 1979).

Chapter 13. Spacemen living on Earth

1. Cynthia Skove, 'Doctor has strange message he wants world to hear', *Daily Press*, Newport News – Hampton, VA, 11 April 1980, reproduced in *UFO Newsclipping Service*, no. 130, p. 10.

2. Carl van Vlierden, 'Investigation under hypnosis: contact revealed', *Flying Saucer Review*, vol. 21 no. 3, p. 10.

3. Horacio Gonzalez Ganteaume, 'Landing at San Pedro de los Altos', *Flying Saucer Review*, vol. 15 no. 2, p. 22.

4. George Adamski, *Flying Saucers Farewell* (1961), published in paperback as *Behind the Flying Saucer Mystery* (Paperback Library, New York, 1967), pp. 62–3.

5. George Adamski, *Inside the Space Ships* (Arco Publishers and Neville Spearman, London, 1956), pp. 31–40.

6. Truman Bethurum, *Aboard a Flying Saucer* (DeVorss & Co., Los Angeles, 1954), pp. 90–4.

7. Orfeo Angelucci, *The Secret of the Saucers* (Amherst Press, WI, 1955), pp. 55–7.

8. Lester J. Rosas, 'The Lester Rosas Story', *Flying Saucers International*, no. 29, pp. 13–15.

9. Howard Menger, *From Outer Space* (Pyramid Books, New York, 1967), pp. 65–7.

10. Stranges' book published 1967 by I.E.C. Inc., Van Nuys, CA.

11. Gordon Melton, 'The Contactees: A Survey', in Mimi Hynek (ed.), *The Spectrum of UFO Research* (J. Allen Hynek Center for UFO Studies, Chicago, 1988), pp. 99–100.

12. UFO Photo Archives, Tucson, Arizona, 1985. A useful summary of the affair was written by Hilary Evans for *The Unexplained* partwork, and can be found in issues nos 134, 135, 137.

13. Jim and Coral Lorenzen, *UFOs Over the Americas* (Signet Books, New York, 1968), pp. 75–7; Charles Bowen, 'More Unusual Humanoids', *Flying Saucer Review*, vol. 14 no. 3, p. 18.

14. Skove, op. cit.; Lou Zinsstag and Timothy Good, *George Adamski – The Untold Story* (Ceti Publications, Beckenham, Kent, 1983), pp. 108–10.

15. 'Spanish Woman Recalls Abduction 36 Years Ago', *Flying Saucer Review*, vol. 29 no. 4, pp. 8–9, compiled from report in Spanish newspaper *El País* of 25 October 1983.
16. Fawcett Crest, Ballantine Books, New York, 1985.

Chapter 14. Is Man alone in the Universe?

1. 'Astronomers Spot Carbon in New Comet', *New Scientist*, 28 May 1987, p. 23, noted in *Science Frontiers*, no. 52, pp. 1–2.
2. Christopher F. Chyba, 'The Cometary Contribution to the Oceans of Primitive Earth', *Nature*, 330:632, 1987, noted in *Science Frontiers*, no. 56, pp. 2–3.
3. R. Monastersky, 'Comet Controversy Caught on Film', *Science News*, 133:340, 1988, noted in *Science Frontiers*, no. 58, p. 1.
4. Report by Lee Dye in *Los Angeles Times*, 2 April 1988.
5. Jonathan Eberhart, 'Have Earth Rocks Gone to Mars?', *Science News*, 135:191, 1989, noted in *Science Frontiers*, no. 63, pp. 2–3.
6. Jeff Hecht, 'Lunar Link with Life on Planets', *New Scientist*, 21 January 1988, p. 40, noted in *Science Frontiers*, no. 56, p. 2.
7. Frank T. Kyte, et al., 'New Evidence on the Size and Possible Effects of a Late Pliocene Oceanic Asteroid Impact', *Science*, 241:63, 1988, noted in *Science Frontiers*, no. 59, p. 3.
8. Wendy S. Wolbach, et al., 'Global Fire at the Cretaceous-Tertiary Boundary', *Nature*, 334:665, 1988, noted in *Science Frontiers*, no. 60, p. 3.
9. Adrian Berry, 'Asteroid theory for late debut of high life-forms', *Daily Telegraph*, 19 February 1990.
10. Adrian Berry, 'Encke: a danger to our planet', *Daily Telegraph*, 24 October 1988.
11. Report by Adrian Berry, *Daily Telegraph*, 21 April 1989.
12. Gail Vines interview with Stephen Jay Gould, 'If only things had been different . . .', *New Scientist*, no. 1705, 24 February 1990, pp. 64–5.
13. See, for example, Paul Davies, *Other Worlds* (J. M. Dent & Sons, London, 1980; Penguin Books, London, 1988).
14. 'A Conversation with a witty UFO', *Northern UFO News*, no. 141, February 1990, pp. 17–18.
15. Carl B. Fliermans and David L. Balkwill, 'Microbial Life in Deep Terrestrial Subsurfaces', *Bio-Science*, 39:370, 1989, noted in *Science Frontiers*, no. 65, p. 3.
16. Don Arnott, 'Micro-organisms and nuclear waste: a neglected problem', SCRAM, 1988.
17. Fred Hoyle and N. C. Wickramasinghe, 'Sunspots and Influenza', *Nature*, 343:304, 1990, noted in *Science Frontiers*, no. 68, p. 3; Fred Hoyle and N. C. Wickramasinghe, *Diseases from Space* (J. M. Dent & Sons, London, 1979; Sphere Books, London, 1981).
18. *Northern UFO News*, as note 14.

Picture Credits

Index